Discover & Learn

Human and Physical

Teacher Book

This Teacher Book accompanies CGP's Discover & Learn Human and Physical Geography books for KS2.

It includes background information to help teachers introduce and teach each topic, answers to Activity Book questions and a range of suggestions for extra activities.

It's the perfect guide to planning and delivering Human and Physical Geography lessons throughout KS2!

Contents

Rivers

The Water Cycle 1
From Source to Sea...............2
A Land Shaped By Rivers........3
Life in a River 4
Exploring the Severn 5
Life Along the Severn 6
Working Along the Severn 7
Meet the Danube! 8
The Danube's History............9
Cities of the Danube 10
Along the Colorado 11
Water Wars........................12
Tourism and the Colorado ...13
The Mighty Amazon..............14
Nature and the Amazon 15
Amazon Under Threat..........16
Be a River Friend 1 17
Be a River Friend 2 18

Volcanoes & Earthquakes

Inside the Earth 19
A Floating Crust 20
Clashes and Collisions............21
Vesuvius 22
Timeline of an Eruption.........23
The Eruption Continues 24
Types of Volcano 25
Volcanoes in the Deep..........26
San Francisco 27
The 1906 Earthquake............28
After the Earthquake 29
Earthquake Information 30
Troubled Earth 31
Keeping Safe – Prediction 32
Keeping Safe – Prevention 33
Keeping Safe – Preparation...34
All Calm in the UK 1 35
All Calm in the UK 2 36

Living Planet

The Solar System...................37
Why Earth?............................38
What Time Is It? 39
More Than Weather............40
Ocean life..............................41
Oceans and the Climate 42
Mountains 43
Tropical Rainforests...............44
Woodlands 45
Hot, Cold and In-Between ...46
Grasslands 47
A Frozen Place......................48
Water Worlds 49
Making Changes....................50
Biome Summary 51
Climate Change 52
Effects of Climate Change.....53
The Future 54

Published by CGP

Consultant: Joanna Copley

Contributors: Mai Black, Julianne Britton, Mark McDermott

Editors: Mary Falkner, Sarah Pattison, Rosa Roberts, Rebecca Russell, Caroline Thomson

With thanks to Felicity Booth, Ellen Burton, Emma Espley, Harriet Foster, Sharon Keeley-Holden and Rachael Rogers for the proofreading.

With thanks to Jan Greenway for the copyright research.

National Curriculum references throughout reproduced under the terms of the Open Government Licence v3.0. http://www.nationalarchives.gov.uk/doc/open-government-licence/version/3/

When using the Extra Activities in this product, please take the safety of the participants into consideration at all times, and ensure that children are supervised when researching material for this product online. Teachers should also take into account pupils' personal circumstances when dealing with topics of a sensitive nature.

ISBN: 978 1 78294 986 2

Printed by Elanders Ltd, Newcastle upon Tyne

Text, design, layout and original illustrations © Coordination Group Publications Ltd. (CGP) 2019

All rights reserved.

Photocopying more than 5% of this book is not permitted, even if you have a CLA licence.
Extra copies are available from CGP with next day delivery • 0800 1712 712 • www.cgpbooks.co.uk

The Water Cycle

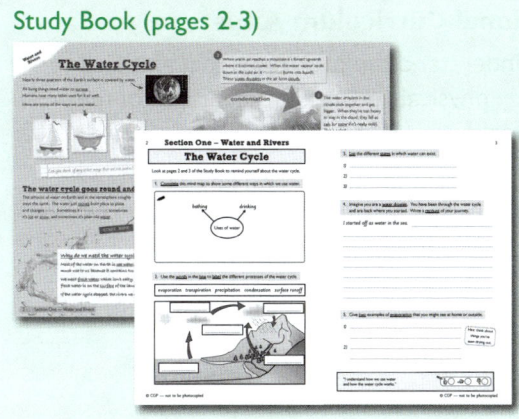

Study Book (pages 2-3)

Activity Book (pages 2-3)

National Curriculum Aims

- Understand the processes that give rise to key physical geographical features of the world.
- Describe and understand key aspects of physical geography, including rivers and the water cycle.

Introduction

This first topic introduces pupils to the water cycle — a fundamental process that keeps the world's rivers flowing and provides us with fresh water. There's a fixed amount of water on Earth, which has been continually circulated through the water cycle for about 4.6 billion years. Only 2.5% of this water is fresh water, and the majority of this is locked up in glaciers and ice sheets, so we rely on the water cycle to keep renewing less than 1% of the water on Earth. The water cycle happens when water evaporates from the ocean — the salt doesn't evaporate and is left behind — creating water vapour, some of which will eventually fall onto the land as rain.

Before pupils complete the Activity Book, get them to consider the importance of the water cycle by asking them what they think might happen if the water cycle suddenly stopped.

Answers to Activity Book Questions

1. E.g. cleaning, transport (i.e. sailing), water sports and swimming.
2. In the centre: transpiration.
   Clockwise from top left: condensation, precipitation, surface runoff, evaporation.
3. Ice/snow, water and water vapour / solid, liquid and gas.
4. Any appropriate answer. Pupils don't have to include all stages of the water cycle as long as they get back to where they started.
5. E.g. clothes drying, puddles drying out, steam from tea/coffee, hair drying.

Extra Activities

- Show pupils an annual precipitation map of the UK. Ask them to identify the most mountainous areas of the UK on the map. Ask pupils if they can see any link between how mountainous an area is and the amount of precipitation it receives.

- Pupils can further explore evaporation and condensation in Science. Get pupils to look at salt water solutions in a saucer and observe how, when left for a few days, the water evaporates and leaves the salt behind. To speed up this process, an oven can be used to heat and evaporate the water. Pupils can see condensation in action by observing what happens when a cold plate is held over a boiling kettle.

- Ask pupils to design a poster to explain the water cycle to others.

- Ask pupils to write a song to help them remember the stages of the water cycle. Alternatively, search online for a song about the water cycle that pupils could learn as a class.

Discover & Learn Human and Physical Geography — Rivers

From Source to Sea

Study Book (pages 4-5)

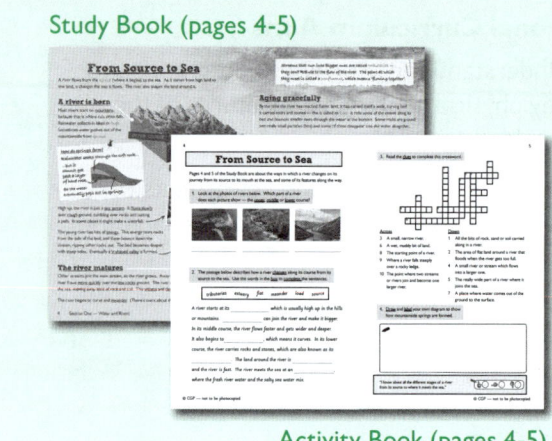

Activity Book (pages 4-5)

National Curriculum Aims
- Understand the processes that give rise to key physical geographical features of the world.
- Describe and understand key aspects of physical geography, including rivers, mountains and the water cycle.

Introduction

Following on from the water cycle, pupils now learn in more detail how rainwater makes its way to the sea. The upper, middle and lower courses of a river have very different characteristics from each other.

As there is a lot of new vocabulary on these pages, it may be helpful to write some key words on the board before starting the activities. Pupils could draw their own flashcards with a picture of a river feature on one side and its name and definition on the other.

Answers to Activity Book Questions

1. Lower, upper, middle.

2. A river starts at its *source* which is usually high up in the hills or mountains. *Tributaries* can join the river and make it bigger. In its middle course, the river flows faster and gets wider and deeper. It also begins to *meander*, which means it curves. In its lower course, the river carries rocks and stones, which are also known as its *load*. The land around the river is *flat* and the river is fast. The river meets the sea at an *estuary*, where the fresh river water and the salty sea water mix.

3. Across: 3 — stream, 6 — bog, 8 — source, 9 — waterfall, 10 — confluence
   Down: 1 — load, 2 — floodplain, 4 — tributary, 5 — estuary, 7 — spring

4. Any appropriate drawing and labels.

Extra Activities

- Split pupils into pairs and ask each pupil to write clues describing a certain part of a river. They should then give their clues to their partner, who has to work out which part of the river the clues describe.

- As a class, read some poems about rivers. E.g. *The River* by Valerie Bloom, *River Journey* by Moira Andrew and *The Immortal River* by David Windle. After reading one or more of these poems, ask pupils to write their own poem about the course of a river.

- Ask pupils to use an atlas to identify some rivers around the world, e.g. the Rhine, the Columbia River, and the Yellow River. Ask them to identify the countries the rivers flow through and their major tributaries.

- Demonstrate to the class how springs form. Cut the top off a water bottle and make holes at various heights up the side of the bottle. Fill the bottle halfway with sand, then make an impermeable layer with modelling clay or play dough. Fill the bottle the rest of the way with sand. At this point, ask pupils to predict what will happen when water is added. Slowly pour water into the bottle until the water emerges from one or more of the 'springs' (holes). Explain that when the water reaches the impermeable layer and is unable to flow downwards, it's forced sideways.

A Land Shaped By Rivers

Study Book (pages 6-7)

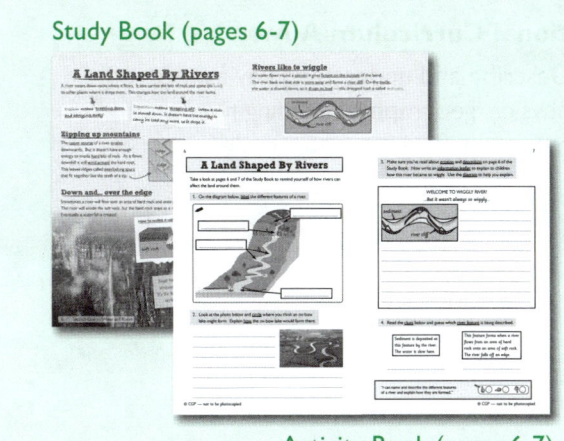

Activity Book (pages 6-7)

National Curriculum Aims

- Understand the processes that give rise to key physical geographical features and how they bring about change over time.
- Describe and understand key aspects of physical geography, including rivers, mountains and the water cycle.
- Use maps, atlases, globes and digital maps to locate the features studied.

Introduction

Rivers create many geographical features through erosion and deposition. This topic allows pupils to learn how powerful rivers are and how they can change the landscape. Where a river bends, the majority of the water is pushed towards the outside of the bend. This increases the river's speed on the outside of the bend, leading to more erosion. Landscapes can also be shaped by the erosion at waterfalls. For example, around 12,000 years ago, Niagara Falls was 11.4 km (7.1 miles) further south than it is today. Constant water flow and repeated freezing and thawing has caused the rocks beneath to erode and the Falls to retreat. It's estimated that the falls won't exist 50,000 years from now.

Answers to Activity Book Questions

1. Top to bottom: Interlocking spurs, waterfall, meanders, delta/estuary

2. Pupils should have circled a meander on the river that has a narrow 'neck'. Pupils' explanations should draw on information from page 7 of the Study Book, explaining that the neck of the meander becomes narrower and narrower, and eventually the river erodes through. The river follows the straight path and the meander is cut off, forming the ox-bow lake.

3. Pupils' answers should demonstrate a basic understanding of erosion and deposition and that the speed of different parts of the river is key in creating meanders. They should identify that the slower moving water on the inside of the bend drops its load, leaving sediment behind, while the faster moving water on the outside of the bend erodes the river bank.

4. Clue 1: Delta/estuary. Clue 2: Waterfall.

Extra Activities

- Provide pupils with a blank map of the world and a list of the world's top ten tallest waterfalls. Ask them to use an atlas to locate each waterfall and mark it in the correct location on their map.

- Pupils can explore the relationship between water speed and erosion. Place a small amount of sand, gravel and several pebbles in a piece of guttering. Using a tap or hose, run water down the guttering and over the sand, gravel and pebbles at a low speed. Only the sand should move with the slow water. Replace the sand and repeat with the water at medium and high speed. The sand and gravel should move with the medium-speed water and all the materials should move when exposed to the fast-moving water. Ask pupils if they think a fast-moving river will erode the land more or less than a slow-moving river. Discuss with pupils the speed of rivers on a bend, and why there is more erosion on the outside of the bend than the inside.

- Show pupils photographs of river features, including waterfalls, meanders, interlocking spurs and deltas. Pupils can give a presentation about the photographs, explaining what the features are and how they're formed.

Life in a River

Study Book (pages 8-9)

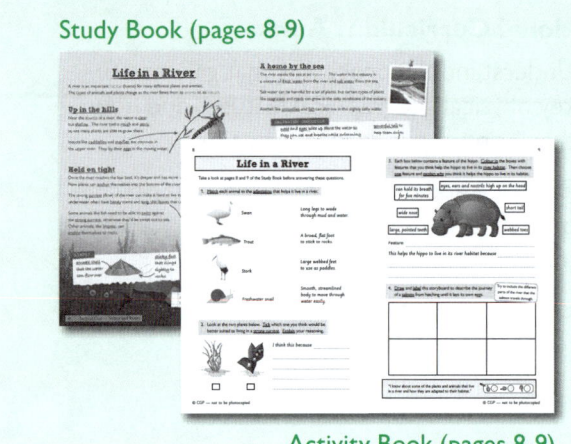

Activity Book (pages 8-9)

National Curriculum Aims
- Describe and understand key aspects of physical geography, including rivers, mountains and the water cycle.

Introduction

As covered on pages 4 and 5 of the Study Book, different parts of a river have different characteristics. These different characteristics create a variety of habitats. The plants and animals that live in the river are adapted to the environment found in that part of the river.

Before reading the Study Book, ask pupils what plants and animals they think might live in or around a river. (Pupils should be able to draw on their existing knowledge of living things and their habitats.) Recap the different features of a river from its source to its mouth and ask the pupils if they think certain plants or animals are better suited to certain locations of the river.

Answers to Activity Book Questions

1. Swan — Large webbed feet to use as paddles, Trout — Smooth, streamlined body to move through water easily, Stork — Long legs to wade through mud and water, Freshwater snail — A broad, flat foot to stick to rocks.

2. Pupils should have ticked the plant with long, thin leaves. Pupils' explanations should demonstrate that they understand that long, thin, flexible leaves help the plant to bend to tolerate a strong current.

3. Pupils should have coloured: can hold its breath for five minutes / eyes, ears and nostrils high up on the head / webbed toes.
   Pupils should have selected one of the above features and identified how it helps the hippo to live in the water, e.g. *Feature:* eyes, ears and nostrils high up on the head. *This helps the hippo live in its river habitat because* it can see, breathe and hear above the water while swimming / when its body is underwater.

4. Any appropriate drawings. Pupils should draw on information from page 9 of the Study Book.

Extra Activities

- Pupils can use what they have learned about life in rivers to create their own imaginary river animal. Ask pupils to describe which part of the river their animal lives in and what adaptations it has to make it well-suited to its particular environment.

- Further to question 4 in the Activity Book, pupils could write a diary entry from the point of view of a salmon, describing the journey back upstream to spawn. To help pupils visualise the journey, search online for a video documenting the return of sockeye salmon from the sea to lay their eggs.

- Ask pupils to research a plant or animal that is found in a river. This can include what part of the river it lives in, and how it's adapted to live in the river. They can then create a presentation about the plant or animal that they have chosen to research.

Discover & Learn Human and Physical Geography — Rivers

Exploring the Severn

Study Book (pages 10-11)

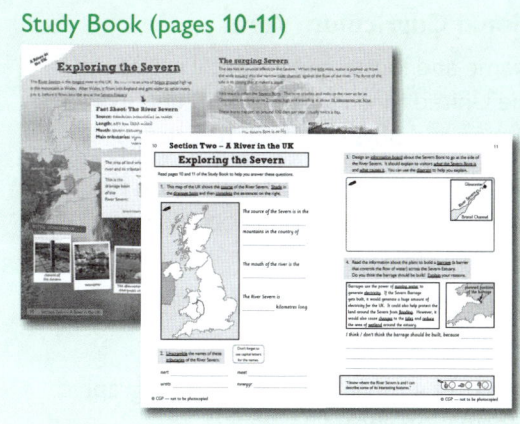

Activity Book (pages 10-11)

National Curriculum Aims
- Name and locate geographical regions of the United Kingdom and their key topographical features, including rivers.
- Describe and understand key aspects of physical geography, including rivers.
- Use maps to describe features studied.

Introduction

This topic gives pupils an overview of the River Severn before they go on to learn about land use and industry around the river in greater detail. The Severn is the longest river in the UK, and the second longest in the British Isles (after the River Shannon in the Republic of Ireland). Its source is a peat bog in Plynlimon in the Cambrian Mountains in Wales, approximately 610 m (2,000 feet) above sea level. Its estuary, where it meets the Bristol Channel, is around fourteen km wide. The Channel joins the Celtic Sea and the Atlantic Ocean.

If possible, show the class a video of the Severn Bore after they have read pages 10 and 11 in the Study Book. This will help them when they come to answer question 3 in the Activity Book.

Answers to Activity Book Questions

1. *The source of the Severn is in the* Cambrian *Mountains, in the country of* Wales. *The mouth of the river is the* Severn Estuary. *The River Severn is* 354 *kilometres long.*
2. nert — Tern,　meet — Teme,　urots — Stour,　nvwyyr — Vyrnwy
3. Any appropriate answer. Pupils should draw on information from page 11 of the Study Book.
4. Pupils may answer either way, as long as they give sensible reasons to support their answer.

Extra Activities

- Provide pupils with a blank map of the River Severn and the surrounding area, as well as photos of points along the river, e.g. the Cambrian Mountains, Tewkesbury, Shrewsbury Cathedral, Iron Bridge, the Forest of Dean, Bristol and the Severn Bridge. The photos should be labelled and numbered. Ask pupils to use a map of the area to mark the number of each photo on the correct location on their own blank map.

- Show pupils a video of the Severn Bore. In groups or as a class, pupils can pretend they are news reporters reporting on the event. They should briefly summarise what causes the bore and include facts such as its height and how far it travelled on that particular day. One or more pupils could pretend to be surfers and the other pupils could interview them to get their thoughts and feelings on the phenomenon.

- The Severn Estuary has been designated as a 'Special Area of Conservation'. Search online for information about a local 'Special Area of Conservation' and provide pupils with some key facts about it. Pupils could then write a speech about why this area needs to be protected. Alternatively, pupils could choose a local piece of land and write a persuasive piece about why they believe it should become a 'Special Area of Conservation'.

Discover & Learn Human and Physical Geography — Rivers

Life Along the Severn

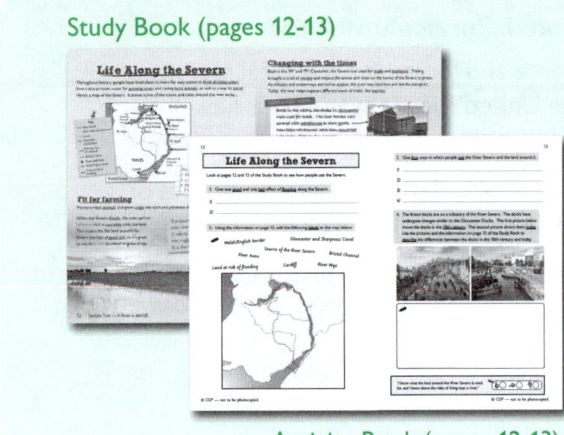

Study Book (pages 12-13)

Activity Book (pages 12-13)

National Curriculum Aims

- Name and locate geographical regions of the United Kingdom and their identifying human and physical characteristics.
- Understand the processes that give rise to key physical and human geographical features and how these are interdependent.
- Understand land-use patterns and how they have changed over time.
- Interpret a range of sources of geographical information, including maps.

Introduction

This topic introduces pupils to different forms of land use around a river, from farming to industry. During the Industrial Revolution, the Severn, like many UK rivers, was used for trade and transport. Many factories were also built along its banks (there's more on factories on page 14 of the Study Book and Activity Book). Waterways were later overtaken by roads, railways and air travel as these provided quicker and easier means of transport. As a result, some ports and docks, including the Gloucester docks, fell into disuse. The Gloucester docks have since been renovated and have been transformed into residential and recreational areas.

Before pupils read the section, recap the ways in which we use water, and ask them to identify any ways in which people use rivers specifically.

Answers to Activity Book Questions

1. E.g. Good: When the river floods, it brings sediment onto the land around the river. The sediment is full of nutrients and makes the soil fertile. This makes the land good for crop farming.
   Bad: Flooding can damage homes and other buildings. / Flooding is dangerous and people can get hurt.
2. *Welsh/English border* — dashed line, *Land at risk of flooding* — hashed shading, *Source of the River Severn* — the river starts in North West Wales. All other labels should match those on page 12 of the Study Book.
3. E.g. Raising farm animals / growing crops / fresh water / trade / transport / water sports / sailing.
4. Pupils' answers should display an understanding that the river is now mostly used for leisure. They may describe there being fewer boats, new buildings, shops, more tourists etc.

Extra Activities

- Ask pupils to research various flood defence methods either in books or online, for example, dams, diversion canals, reservoirs, metal barriers along river banks, and sand bags. Divide pupils into groups and ask each group to research one defence method in detail. Each group can then present an overview and some pros and cons of their flood defence method to the rest of the class.
- Share images and video clips of news footage (available online) of the 2007 flooding of the River Severn with the class and provide pupils with key facts about the event. Ask pupils to write a newspaper report about the flooding of the River Severn.
- Provide pupils with a map of the River Severn, along with images of tourist attractions along the river. The images could include Ironbridge, Bridgenorth Cliff Railway, Worcester Cathedral and the Gloucester Quays. Ask pupils to imagine that they've been on a boat travelling along the Severn. Get them to write a postcard describing their journey, including some of the places they've seen from the boat.

Discover & Learn Human and Physical Geography — Rivers

Working Along the Severn

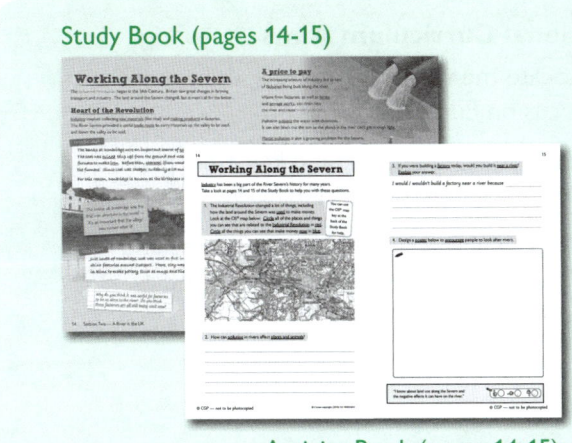

Study Book (pages 14-15)

Activity Book (pages 14-15)

National Curriculum Aims
- Understand the processes that give rise to key physical and human geographical features and how these are interdependent.
- Understand land-use patterns and how they have changed over time.
- Describe and understand key aspects of human geography, including land use and economic activity including trade links.

Introduction

Industry along the Severn has changed a great deal since the Industrial Revolution. The Severn was a vital transport link between the coalfields and the Bristol Channel. Ironworks, brick and tile works and porcelain factories all grew along the banks of the Severn throughout the 18th and 19th Centuries. Although polluting industries, like ironworks, no longer operate along the Severn, pollution is still a problem, particularly plastic pollution.

This topic provides an opportunity to discuss the positive and negative impacts of industry, as well as the wider issues of the environment and recycling.

Answers to Activity Book Questions

1. Pupils could circle: Industrial Revolution: Ironbridge / Coalport / Furnaces / Tar Tunnel
   Now: Craft Centre / Museum / Discovery Centre / Caravan Park / Power Station / Hotel

2. Any appropriate answer. Pupils may expand on information in the Study Book to draw their own conclusions. E.g. Pollution in rivers can harm the plants and animals that live in and around the river. Too much pollution can block the sun from shining through the water so plants can't survive. Plastic and chemical waste can kill animals if they eat it.

3. Pupils may answer either way, as long as they give sensible reasons to support their answer.

4. Any appropriate drawing and text. Pupils can draw on information from the Study Book.

Extra Activities

- Show pupils pictures of The Iron Bridge in Ironbridge. It was the first major bridge in the world to be made from cast iron. Discuss other materials that bridges can be made from and ask pupils why they think building a bridge from iron is a good idea (e.g. it's very strong and lasts a long time).

- Show pupils pictures or a video of waterwheels. As a class, discuss why the water makes the wheel turn and why this is useful for industry (e.g. turning mill wheels to grind flour). Pupils can then make their own waterwheels — this can be done using two paper plates placed about 5 cm apart, with a stick through the centres of both plates. Plastic cups should be attached between the two plates around the edge, all facing the same way. Pour water from above to fill each cup in turn. This should make the wheel rotate with the weight of the water. Pupils can also look at how a stream flowing under the wheel can cause it to turn.

- Take pupils to a local waterway. Ask them to make a list of the different types of litter along the river that they can see. Back in the classroom, ask them to discuss the effects that pollution and littering could have on the environment. They can also write a letter to their local council about the pollution they saw. Alternatively, contact the local council to organise a litter pick for the class along a local river bank.

Meet the Danube!

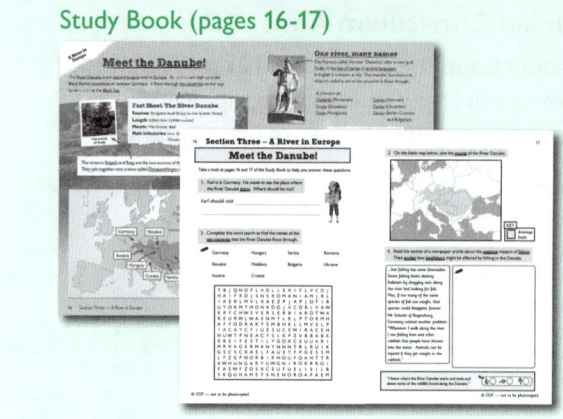

Study Book (pages 16-17)

Activity Book (pages 16-17)

National Curriculum Aims
- Locate the world's countries, using maps to focus on Europe.
- Describe and understand key aspects of physical geography, including rivers.
- Understand geographical similarities and differences of a region of the UK and a region in a European country.

Introduction

This topic introduces pupils to the River Danube. The river itself passes through ten Western, Central and Eastern European countries, but its drainage basin spans nineteen. Its delta forms the largest wetland in Europe and stretches across parts of Ukraine and Romania. The delta is a UNESCO World Heritage site, due to its impressive biodiversity.

Once pupils have read pages 16-17 of the Study Book, recap the physical characteristics of rivers and ask them if they can identify similarities and differences between the Danube and smaller rivers like the Severn.

Answers to Activity Book Questions

1. *Karl should visit* the Black Forest (near Donaueschingen).
2. On the right are the locations of the ten countries in the grid:
3. Pupils should have plotted the correct course of the river, following the information from the Study Book.
4. Any appropriate answer. Pupils should draw on information from the Study Book and Activity Book. E.g. The kingfisher could get caught and injured in old fishing lines when it's catching fish. / If too many fish are caught by fishermen, the kingfisher could run out of food.

Extra Activities

- Ask pupils to research the effects of the 2013 Danube flood on the town of Passau, Germany. In this town, two tributaries join the Danube, which causes regular flooding. In 2013 the flood reached over three metres on the wall of the town hall. They could then write a newspaper report about the event.

- Provide pupils with blank maps of the countries of the Danube. Tell them that languages spoken in different places can be grouped into language families or branches. In the Danube basin, four groups can be found — Germanic (German and Austrian), Urgic (Hungary), Romantic (Romanian, spoken in Romania and Maldova), and Slavic (Slovak, Ukrainian, Bulgarian, Serbian and Croatian). Ask pupils to colour their maps with the language group that is spoken in each country and to create a key. Discuss with pupils why some countries might have similar languages. Languages groups have common root languages (e.g. Romanian is descended from Latin), and these languages spread and changed as people have moved around Europe.

- Recap what pupils already know about simple food chains. Ask them how they think one type of animal can affect another. Discuss the ways in which human actions like overfishing can indirectly harm other animals.

The Danube's History

Study Book (pages 18-19)

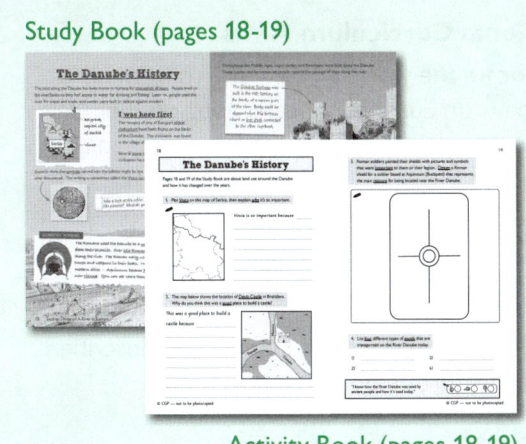

Activity Book (pages 18-19)

National Curriculum Aims

- Describe and understand key aspects of physical geography, including rivers.
- Describe and understand key aspects of human geography, including land use, and economic activity including trade links.
- Identify key topographical features (including rivers), and land-use patterns; and understand how some of these aspects have changed over time.

Introduction

This topic gives pupils the opportunity to consider land use around a river in a modern and historical context. The Danube was the home of ancient civilisations such as the Vinca and the Romans. In the years 4000-5000 BC, the Vinca culture inhabited Serbia and parts of Bulgaria and Romania. They used the Danube and the land around the river to support their fishing and agricultural lifestyles. By the 1st century AD, the Danube formed part of the northern border between the Roman Empire and their Germanic neighbours.

Once pupils have read pages 18-19 of the Study Book, discuss how land use has changed over the past 7000 years. Can pupils think of any reasons why we don't all need to live near rivers anymore?

Answers to Activity Book Questions

1. E.g. *Vinca is so important because* it is the site of one of the oldest civilisations in Europe and where what might be the oldest form of writing was discovered.

2. Any appropriate answer. Pupils should draw on information on page 19 of the Study Book. E.g. *This was a good place to build a castle because* it is near the Danube and the Morova River so they could control passing ships on both rivers. / the Danube is a natural barrier that protects them from their enemies.

3. Any appropriate drawing. Pupils should draw on information from the Study Book (e.g. a castle, boats, people swimming).

4. Cereals, fruits, vegetables, animals.

Extra Activities

- As a link to the KS2 History Romans topic, ask pupils to use their shield designs on question 3 in the Activity Book to make A4 size models of their shields. These should be made from cardboard and painted with their designs. They could use papier maché to make the shields 3D. Once the shields are dried, they could be displayed on the wall or made into a display to show how the Romans would use them in battle. (They would have used a formation where first row of soldiers held their shields in a wall at the front of the legion, and the other rows would hold their shields up over their heads to form a shield roof.)

- Give pupils a map of the Danube as it flows through the north of Austria. The map should include the following castles on the banks of the river: Castle Greinburg, Werfenstein Castle, Schloss Persenbeug, Schloss Schönbühel, Ruine Hinterhaus, and Burgruine Dürnstein. Add a 12 x 12 grid on top of the map with the squares labelled 1-12 along one axis, and A-L along the other (for older pupils, lines on both axes should be labelled as 1-12). Provide pupils with grid references for each castle and ask them to add the OS® symbol for fortress/castle to the map in the correct places. Once pupils have marked on all the castles, ask pupils to shade in all the squares they think would make a good location for another castle.

Cities of the Danube

Study Book (pages 20-21)

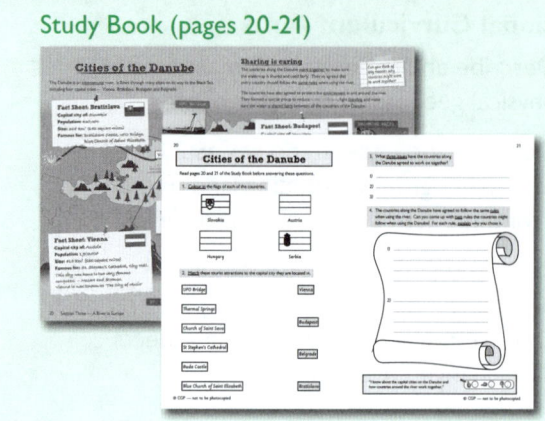

Activity Book (pages 20-21)

National Curriculum Aims
- Locate the world's countries, using maps to focus on Europe, concentrating on their key human characteristics, countries and major cities.
- Describe and understand key aspects of human geography including types of settlement and land use.
- Use maps to locate countries and describe the features studied.

Introduction

The Danube runs through four capital cities and many other cities. Of the capital cities the Danube passes through, Belgrade and Bratislava are the oldest settlements. A settlement in the area of modern-day Belgrade was part of the Vinca culture in around 4000 BC. The first permanent settlement in Bratislava was during the Linear Pottery Culture, around 5000 BC. A Celtic settlement was later built in Bratislava, as were settlements in Vienna and Budapest, which were later invaded by the Romans.

After pupils have read pages 20-21 of the Study Book, ask them why they think so many cities have developed on the banks of the Danube.

Answers to Activity Book Questions

1. Appropriate colouring that matches the flags show on Activity Book pages 20 and 21.
2. UFO Bridge — Bratislava, Thermal Springs — Budapest, Church of Saint Sava — Belgrade, St Stephen's Cathedral — Vienna, Buda Castle — Budapest, Blue Church of Saint Elizabeth — Bratislava
3. Reducing water pollution, fighting flooding and sharing water fairly.
4. Any appropriate answer. Pupils can draw on information from the Study Book or think up their own rules (e.g. factories must not pollute the river), as long as they give sensible reasons to support their answer.

Extra Activities

- Ask pupils to write a non-chronological report about one of the cities featured on pages 20-21. This could include information about famous landmarks, its physical geography, climate, size, culture, famous foods, festivals etc.

- As a class, listen to Strauss's *The Blue Danube*. Ask the pupils how the music makes them feel. Does it sound like a happy or sad piece? Can they identify any of the instruments they can hear? Does any part of the piece make them think of a river? Ask them to listen to another piece of music composed about a river and compare it to the *The Blue Danube*. Some suggestions include: Whiteacre's *The River Cam*, Smetana *The Moldau*, Handel's *Water music*, and *Yellow River Piano Concerto* based on the *Yellow River Cantata* by Xian Xinghai. Ask pupils to think of key words to describe each piece.

- Discuss with pupils why some locations are better than others for settlements (e.g. defendable, source of water, transport). Then ask pupils to imagine they are in charge of building a new city. Would they build it on the banks of a river? If not, why not? Pupils could draw and label a map of their imaginary city and the surrounding landscape, explaining why they chose to build it where they did.

Along the Colorado

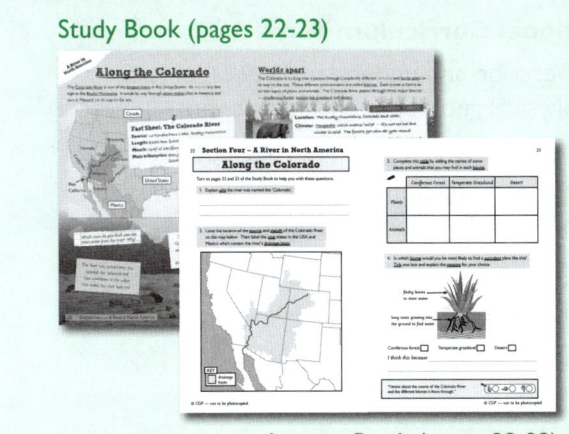

Study Book (pages 22-23)

Activity Book (pages 22-23)

National Curriculum Aims

- Use maps to locate North America, concentrating on its environmental regions and key physical characteristics.
- Describe and understand key aspects of physical geography, including rivers.
- Describe and understand biomes.

Introduction

This first topic will introduce pupils to the Colorado River, its physical geography and some of the diverse biomes it flows through. The Colorado River is perhaps most famous for its role in carving out the Grand Canyon, which is located in Arizona. However, the Grand Canyon only surrounds 446 km of the river's total 2330 km length. The rest of the river's course is very diverse, flowing through mountains and forests as well as valleys and deserts.

After pupils have read pages 22 and 23 of the Study Book, compare pictures of the Colorado River as it flows through the Rocky Mountains and the Mojave Desert and discuss all the differences they can see.

Answers to Activity Book Questions

1. E.g. The Colorado was named after the Spanish word 'colorado', meaning coloured red, because red sandstone made the river look red.

2. Pupils' labels for the states should match the map on page 22 of the Study Book. They should also have correctly labelled the source as La Poudre Pass Lake in the Rocky Mountains, and the mouth as the Gulf of California, Mexico.

3. *Coniferous forest plants*: e.g. pines, spruces, Christmas trees. *Coniferous forest animals*: e.g. deer, elk, black bears, cougars. *Temperate grassland plants*: e.g. tall grasses. *Temperate grassland animals*: e.g. bison, prairie dogs, coyotes, hawks, owls. *Desert plants*: e.g. cacti. *Desert animals*: e.g. jaguars, tarantulas, scorpions, vultures

4. Pupils should have ticked: Desert E.g. *I think this because* the plant has long roots and fleshy leaves that help it to find and store water. / it's adapted to survive in a biome with very little rainfall.

Extra Activities

- Provide pupils with a map of the Colorado River and its drainage basin and a basic map of the biomes in the US. Pupils can compare the maps and shade in different areas of the drainage basin to indicate the type of biome in that location. Pupils should add a key to indicate which biome each colour signifies.

- Provide pupils with a map of the Colorado River and add a numbered grid on top of it. Add dots to indicate the locations of major cities along the river — e.g. Moab, Page, Yuma, San Luis Río Colorado, Bullhead City, Grand Junction and Glenwood Springs. Give pupils a list of the city names with their coordinates from the grid on the map they were given, and ask them to match the cities to the dots on the map.

- Using what they have learnt about biomes, ask pupils to present a weather forecast for a day in either summer or winter in an area surrounding the Colorado River.

Water Wars

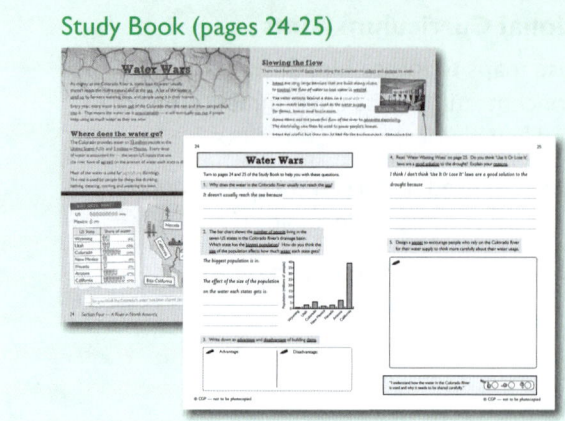

Study Book (pages 24-25)

Activity Book (pages 24-25)

National Curriculum Aims

- Describe and understand key aspects of physical geography, including rivers.
- Describe and understand key aspects of human geography, including the distribution of natural resources such as water.
- Describe and understand key aspects of human geography, including land use.

Introduction

Until now, pupils may have assumed that the water cycle ensured that rivers had an endless supply of water. However, this topic introduces pupils to a river that cannot support the demand on its water supply.

The Colorado River gets most of its water from melting snow in the Rocky Mountains, but there has been a drought in the southwest of the USA since 1999, which has led to less snow. Coupled with the massive demand for water for agricultural and domestic use, water supplies in the reservoirs along the Colorado are depleting. The Colorado River has reached the sea only a few times since the 1960s. This has led to strict policies on water use, although some of these may not improve the situation. The 'use it or lose it' laws are based on those written over 100 years ago, and in total allocate more water than the river can provide.

Answers to Activity Book Questions

1. E.g. *It doesn't usually reach the sea because* almost all of the water is used up by agriculture and by people.
2. *The biggest population is in* California. E.g. *The effect of the size of the population on the water each state gets is* that the more people live in each state, the more water the state gets.
3. Advantages: e.g. Dams create reservoirs which supply drinking water. / Dams can be used to generate electricity. / Dams can be used to control the flow of water so less is wasted.
   Disadvantages: e.g. Dams change the way the water flows, which can change or destroy animals' habitats.
4. Any appropriate answer. Pupils should draw on information from the Study Book.
5. Any appropriate answer. Pupils may include reasons why using too much water could cause problems, and suggestions for how to save water.

Extra Activities

- Ask pupils to present the data from the 'Share of Water' pictogram on page 24 of the Study Book in a different way, e.g. bar chart, pie chart, bar line graph, etc.
- Get pupils to look at a UK dam or reservoir and research the impact it has had on the local area. This could include looking at before and after pictures (e.g. of West End in North Yorkshire and Derwent in Derbyshire — settlements that were levelled and submerged to make way for the Thruscross and Kielder Water reservoirs respectively) and thinking about the positive and negative impacts for people and wildlife. Ask pupils to imagine that they are attending a town hall meeting where members of the community are debating whether or not to build a dam nearby. Pupils could play the roles of scientists, farmers and homeowners and give their opinions on the dam.
- As a class, discuss and share ideas about how your school could save water.

Tourism and the Colorado

Study Book (pages 26-27)

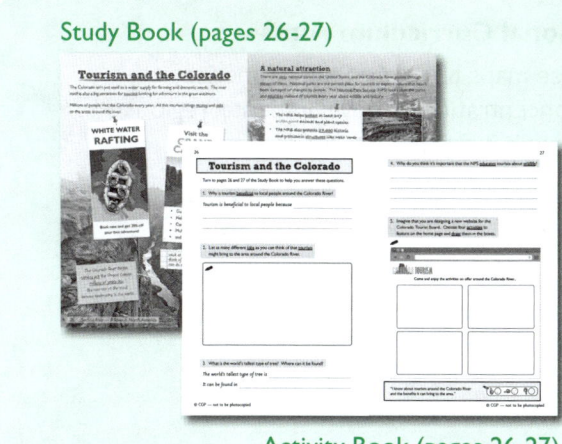

Activity Book (pages 26-27)

National Curriculum Aims
- Use maps to locate North America, concentrating on key physical and human characteristics.
- Describe and understand key aspects of human geography, including economic activity.
- Use maps, atlases and digital mapping to locate states in the USA and describe features studied.

Introduction

The Grand Canyon is the most popular attraction on the Colorado. It attracts over 6 million visitors every year. They come to enjoy the views, tour the area and take part in water sports on the river. Tourism in the Grand Canyon National Park generates over $500 million annually and supports nearly 10,000 jobs in the surrounding areas. However, the volume of visitors brings its share of negative impacts. Helicopter tours of the Canyon create noise pollution and disturb native animals. Millions of visitors following the same trails leads to erosion of the land and destruction of plants and ecosystems. Traffic congestion and littering are also significant issues.

Answers to Activity Book Questions

1. *Tourism is beneficial to local people because* it brings money and jobs to the area.
2. Any appropriate suggestions. E.g. Tour guide, hotel porter/cleaner/manager, restaurant owner, chef, server, rafting instructor, park ranger, gift shop owner, taxi driver.
3. *The world's tallest type of tree is* the California Redwood.
   It can be found in Redwood National Park, California.
4. Any appropriate answer. Pupils should draw on information from the Study Book. E.g. educating people about wildlife may encourage them to think about their impact on the environment or get involved in helping to protect it.
5. Any appropriate drawings. Pupils should draw on information from the Study Book. E.g. people rafting, helicopter rides, canoeing, watching wildlife, hiking, etc.

Extra Activities

- Ask pupils to research one of the National Parks that the Colorado River passes though. They should then imagine that they are on holiday in the park and create a postcard to send home, describing what they have seen and done.
- After reading pages 26 and 27 of the Study Book, ask pupils to work as a class to list positive and negative effects that tourism can have. Hold a class debate about the benefits and drawbacks of tourism around the Colorado River.
- Pupils could plan a trip to the following places: Grand Canyon, Rocky Mountains National Park, Glenwood Springs, Las Vegas, San Francisco, Redwood National Park. Ask them to first locate these places on a map, then consider how they would travel between locations, where they might stay and what route they might take. More advanced pupils could use the scale of the map and an average travel speed (e.g. 70 mph by car) to work out how long it would take to travel between places.

Discover & Learn Human and Physical Geography — Rivers

The Mighty Amazon

Study Book (pages 28-29)

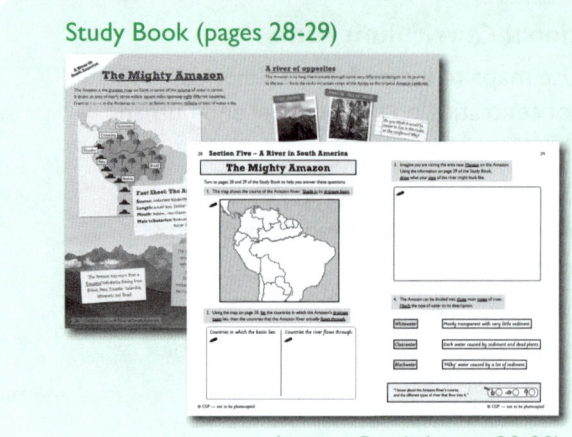

Activity Book (pages 28-29)

National Curriculum Aims
- Use maps to locate South America, concentrating on environmental regions and key physical characteristics.
- Describe and understand key aspects of physical geography, including rivers.
- Describe and understand key aspects of human geography, including the distribution of natural resources such as water.

Introduction

This topic introduces pupils to the Amazon River. The Amazon River basin has an area of around 7 million square kilometres (2.7 million square miles) — so large that the UK could fit inside it more than 28 times. The basin contains a number of biomes, from alpine tundras to savannahs, but the largest of all is the tropical rainforest biome.

Due to extensive debate over the location of the source and mouth of the river, it has been suggested that the Amazon may in fact be the longest river in the world. However, the Nile is currently still widely considered to hold that title. Once pupils have read pages 28 and 29 of the Study Book, ask pupils if they know anything about the Amazon River. What do they think the weather is like? Would it be the same in the Andes as in the rainforest? What kinds of plants and animals do they think might live in or along the river?

Answers to Activity Book Questions

1. Pupils' drawings should match the map on page 28 of the Study Book.
2. Countries in which the basin lies: Venezuela, Colombia, Ecuador, Peru, Bolivia, Brazil.
   Countries the river flows through: Peru, Brazil.
3. Any appropriate drawing. Pupils' drawings could represent the Meeting of Waters or the rainforest setting.
4. Whitewater — 'Milky' water caused by a lot of sediment.
   Clearwater — Mostly transparent with very little sediment.
   Blackwater — Dark water caused by sediment and dead plants.

Extra Activities

- Provide pupils with photographs of either the Andes or the Amazon rainforest. Pupils could develop their descriptive writing skills by writing a detailed description of one of the places in the pictures.

- Give pupils a printed map of South America. Ask them to use the information in the Study Book alongside an atlas to add various labels to their map. For example, you could ask them to label countries, capital cities, rivers, mountain ranges etc. The number and type of labels could be adapted to suit the ability of the pupils.

- Pupils could read further information about the Meeting of Waters or watch an online video explaining the phenomenon. Pupils can then explore mixtures using a variety of different liquids e.g. milk, water, cooking oil, coffee, honey, liquid soap. Pupils could explore what happens when different liquids are mixed. Do they become one liquid or stay separate? Do they look different once mixed?

Discover & Learn Human and Physical Geography — Rivers

Nature and the Amazon

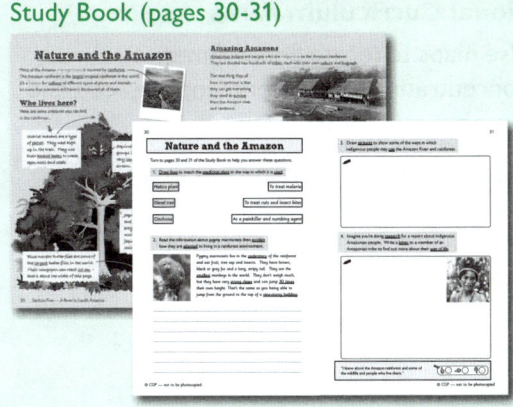

Study Book (pages 30-31)

Activity Book (pages 30-31)

National Curriculum Aims

- Use maps to locate South America, concentrating on environmental regions and key physical characteristics.

- Describe and understand key aspects of physical geography, including biomes and rivers.

- Describe and understand key aspects of human geography, including types of settlement and land use.

Introduction

The Amazon rainforest is the largest tropical rainforest in the world and produces over 20% of the world's oxygen. It also contains about 10% of all the world's known species. It's estimated that a new species is discovered in the Amazon rainforest every three days.

As well as looking at the huge biodiversity in the Amazon rainforest, this topic also provides pupils with an opportunity to consider different lifestyles by thinking about indigenous cultures and how they use the natural resources of the rainforest to survive. Once pupils have read pages 30 and 31 of the Study Book, ask them if they think they could survive in a forest in the UK like the indigenous people do in the Amazon rainforest. If not, why not? What differences are there between a forest in the UK and the Amazon rainforest?

Answers to Activity Book Questions

1. Matico plant — As a painkiller and numbing agent. Diesel tree — To treat cuts and insect bites. Cinchona — To treat malaria.

2. E.g. Pygmy marmosets can jump very high which helps them jump through trees.
   Their strong claws stop them falling out of trees.
   Their dark fur helps them blend in to the dark, shady rainforest.
   They eat food that is readily available in the trees where they live.
   They are light so they can climb on very thin branches to reach their food.

3. Any appropriate drawings. Uses of the river could include: transport, fishing and washing clothes. Uses of the rainforest could include chopping down trees in order to build wooden boats, bridges and houses, making clothing, picking or growing food and collecting plants for medicines.

4. Any appropriate answer.

Extra Activities

- Pupils can explore artwork focused on the Amazon rainforest. They could look at the 'Spirit of the Rainforest' art project created by the Eden Project (a virtual tour of the artwork is available online). Pupils could use the artwork from this project as inspiration to create their own rainforest image.

- Provide pupils with further information about the animals of the Amazon rainforest through books and websites. Ask pupils to choose an animal and create a fact file for a class book on Amazonian animals.

- Ask pupils to produce a flyer informing people about the uses of some of the plants growing in the Amazon rainforest, e.g. the Matico plant, Diesel Trees, rubber trees, cacao, banana trees, etc.

Amazon Under Threat

Study Book (pages 32-33)

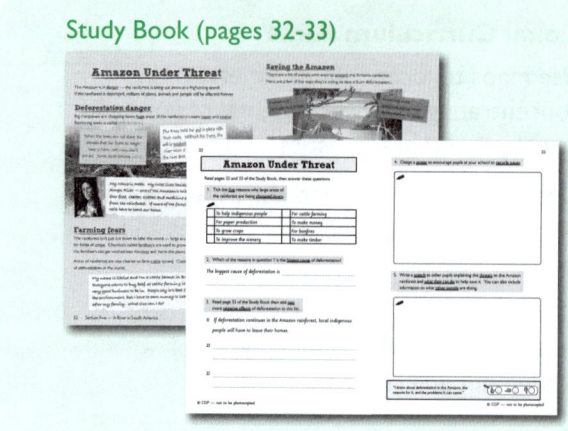

Activity Book (pages 32-33)

National Curriculum Aims
- Use maps to locate South America, concentrating on environmental regions and key physical characteristics.
- Describe and understand key aspects of human geography, including land use, economic activity and the distribution of natural resources.

Introduction

We rely on farming in the Amazon River basin for many of the foods we consume daily. Palm oil (which is present in around half of all supermarket products), fruits, nuts, vegetables, chocolate and coffee are all grown in this area. However, this often comes at the expense of the rainforest. This topic gives pupils the opportunity to consider the positive and negative effects of farming in the Amazon rainforest, as well as considering the ways in which small actions like recycling can have a positive impact on the future of the rainforest.

Answers to Activity Book Questions

1. Pupils should have ticked: For paper production, To grow crops, For cattle farming, To make money, To make timber.

2. *The biggest cause of deforestation is* cattle farming.

3. Any two from: Fertilisers get washed into the river when it rains, which can be harmful to plants and animals. / Cutting or burning down trees destroys habitats for many plants and animals. Some could go extinct. / Soil gets washed into the river without the trees to hold it in place. This can cause the river bed to rise and the land to flood.

4. Any appropriate drawing. Pupils should draw on information from the Study Book.

5. Any appropriate answer. Pupils should draw on information from the Study Book.

Extra Activities

- Hold a class debate about farming in the Amazon rainforest. Half of the pupils represent the local indigenous people like Mahí. The other half represent Brazilian farmers like Carlos.

- Provide pupils with data regarding deforestation in the Amazon rainforest over time, such as the percentage of the Amazon rainforest remaining in Brazil (this data can be found online). Pupils can create graphs using these figures.

- Get pupils to make their own recycled paper. Pupils will require: scrap paper, a piece of very fine wire or plastic mesh, a shallow basin and hot water. Soak torn up pieces of scrap paper in hot water for a few hours, then place in a blender. Blend the paper until it turns to pulp. Add more water to the blender if the pulp doesn't reach pouring consistency. Pour a few centimetres of water into a shallow basin and place the screen into the basin. Pour the pulp into the water, spreading it out evenly. Lift the screen out of the water, leaving a thin layer of pulp across the screen. Allow to drip dry. After a few minutes, turn the mesh upside down on top of a towel and peel off the paper. Once the paper has fully dried, it is ready to use.

Be a River Friend 1

Study Book (pages 34-35)

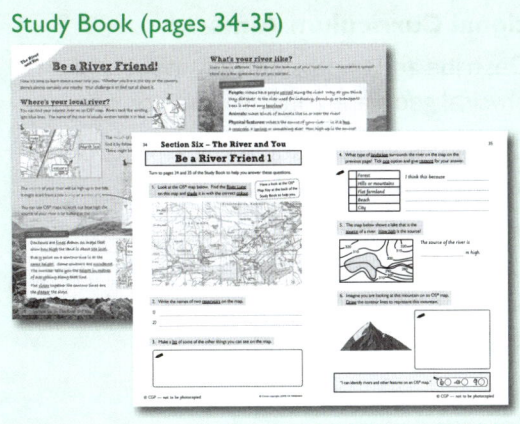

Activity Book (pages 34-35)

National Curriculum Aims
- Describe and understand key aspects of physical geography, including rivers.
- Describe and understand key aspects of human geography, including types of settlement and land use.
- Use fieldwork to observe, record and present the human and physical features in the local area.

Introduction

This final topic gives pupils the opportunity to familiarise themselves with OS® maps and how geographical features like rivers and hills are represented on them.

After pupils have read pages 34 and 35 of the Study Book, recap the various features of rivers that pupils learned about in Section One of the Study Book, and how rivers change as they flow from source to mouth.

Answers to Activity Book Questions

1. Pupils should have shaded the River Lune from the bottom left corner of the map, through Selset and Grassholme Reservoirs and then up towards the upper right corner of the map.
2. Selset Reservoir and Grassholme Reservoir.
3. E.g. trees, phone, nature reserve, visitor centre, camping and caravan site, picnic site, buildings.
4. Pupils should have ticked: hills or mountains. Pupils' explanations should demonstrate that they have identified the contour lines on the map, which show that this is an upland area.
5. *The source of the river is* 300 *m high*
6. Pupils' drawings should demonstrate that they understand that one side of the mountain is steeper than the other, and consequently, the contour lines on that side of the mountain should be closer together.

Extra Activities

- Pupils could make their own flashcard game to help them remember the meaning of each OS® symbol. Give each pupil a set of blank white cards and ask them to draw a symbol on one side of the card and write its meaning on the other side. In pairs, pupils should take it in turns to show their partner a symbol. If they give the correct meaning of the symbol, they get to keep the card. The game should continue for about five to ten minutes, depending on the pupils' abilities. The player with the most cards at the end of the game is the winner.

- Demonstrate to the class how to measure distances on a map using a map scale — this can be done by measuring the distance on a piece of string and then comparing the length of the string to the map scale. In pairs, get pupils to take turns to estimate distances between points on a map using the map scale, then measure the distance.

- Explain four-figure grid references on maps using what the pupils know about coordinates. Provide pupils with a simple grid and instructions. Referring to the OS® legend, ask pupils to add various features to the map, e.g. car park, caravan site, beacon, wind turbine, place of worship, hospital, etc. Provide pupils with a scale for the grid and ask them to work out the distances between the features they have plotted.

Be a River Friend 2

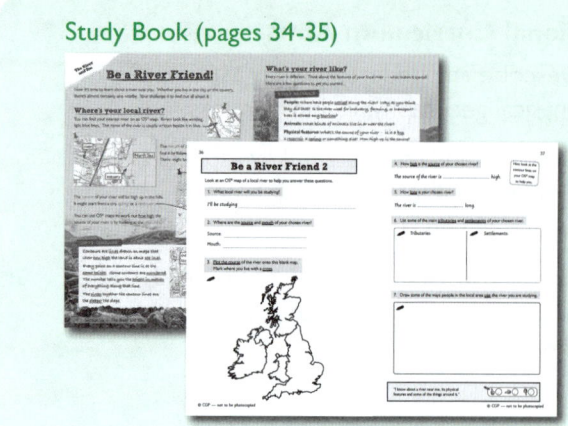

Study Book (pages 34-35)

Activity Book (pages 36-37)

National Curriculum Aims

- Describe and understand key aspects of physical geography, including rivers.
- Describe and understand key aspects of human geography, including types of settlement and land use.
- Use fieldwork to observe, measure, record and present the human and physical features in the local area.

Introduction

The second part of this section allows pupils to research a local river. Provide pupils with an OS® map of a large river that's local to them. If there is no appropriate river nearby, pupils could study a larger river that's further afield, for example the River Thames, Trent, Great Ouse, Wye or Clyde. Examine the physical and human geographical features around the river with them, using the OS® map key, or the one provided in the back of the Study Book.

Pupils could also visit their local river and examine it first-hand. Before doing so, discuss the potential risks of taking a trip to a river. What dangers should pupils be aware of? What should and shouldn't they do around the river? Does the time of year affect what they should be careful of? What about the weather?

Once pupils have studied their river using an OS® map, and possibly a trip to the river, they can report their findings in the Activity Book.

Answers to Activity Book Questions

Answers on these pages will depend on the river studied.

Extra Activities

- If possible, take pupils on a trip to a local river looking for evidence of wildlife. Before the trip, ask pupils what kinds of plants and animals they think live in or near the river. Provide pupils with a checklist of things you'd like them to look for during the visit. Some plants they could search for include: reeds, duckweed, pondweed, hemlock, water forget-me-not and water-lilies. Some animals include: water voles, otters, herons, mallards, swans, kingfishers, mayflies, dragonflies, water boatmen and butterflies. Pupils could be provided with a simple diagram of the section of river they are visiting, so that they can mark on the locations of where they encounter the items on the checklist.

- Pupils could take nets, white plastic trays and magnifying glasses to their local river to fish for invertebrates. Instruct pupils to gently swirl the nets in the water, slowly lift the nets out and turn them inside out into their plastic tray. Using a magnifying glass, pupils can examine what they have caught, before carefully returning everything they caught back into the river.

- Ask pupils to produce a tourist information poster about activities to do along the river, information about how to look after the environment or tips on how to stay safe. Pupils could be provided with tourist leaflets, website print-outs and books to base their information on.

Inside the Earth

Study Book (pages 2-3)

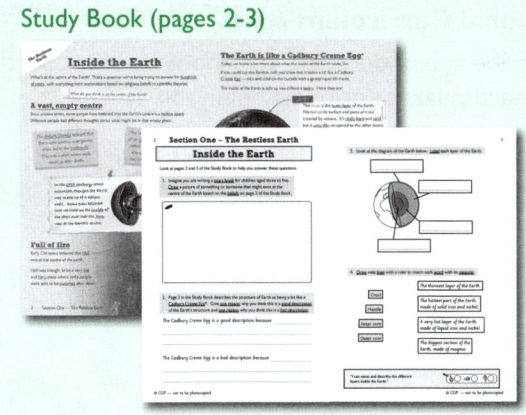

Activity Book (pages 2-3)

National Curriculum Aims

Describe and understand volcanoes and earthquakes, including being able to:

- describe key aspects of the Earth's structure,
- understand that scientific methods help us to answer questions.

Introduction

At different times and in different cultures, there have been many myths about what's at the centre of the Earth. Besides the myths described in the Study Book, another belief in Native American cultures was that the Earth is a great island resting on the back of a giant turtle. Before teaching this topic, it may be helpful to get the pupils to think about what they believe could be at the centre of the Earth and why they think this is.

We know more about what is at the Earth's centre now because of earthquakes. When tectonic plates move, they create seismic waves that travel throughout the Earth's layers. The speed of these waves changes depending on what type of material they're travelling through, so scientists can record them and trace their journeys to determine the structure of the Earth's centre.

Answers to Activity Book Questions

1. Any appropriate drawing based on one of the beliefs given on page 2 of the Study Book.
2. E.g. *The Cadbury Creme Egg® is a good description because* the Earth has a solid outside layer and a liquid middle.
   E.g. *The Cadbury Creme Egg is a bad description because* a Cadbury Creme Egg isn't spherical / the middle bit isn't hot / the very centre isn't solid.
3. *From top to bottom:* Crust, Mantle, Outer Core, Inner Core
4. Crust — The thinnest layer of the Earth.
   Mantle — The biggest section of the Earth, made of magma.
   Inner Core — The hottest part of the Earth, made of solid iron and nickel.
   Outer Core — A very hot layer of the Earth, made of liquid iron and nickel.

Extra Activities

- Ask pupils if they think it would be possible to dig through the centre of the Earth and come out the other side. Ask them to write down their answer and justify it using information from p.2-3 of the Study Book.

- Pupils could use the information on page 3 of the Study Book to write a letter to someone who lived a long time ago to explain the Earth's structure. For example, they could write to someone from Ancient Greece, such as Aristotle — he was interested in the shape of the Earth and wanted to use maths and science to prove it was spherical.

- Pupils could create their own scale model of the Earth's layers in 3D with modelling clay. Using different colours for the different layers, roll a ball for the inner core, then wrap this in another colour to represent the outer core. Do the same for the mantle and crust, keeping the layers roughly the right thickness for the scale. Cover this in blue modelling clay and add some green pieces as land masses. Finally, slice through the sphere to show all the layers. If modelling clay isn't available, pupils could make a 2D version from card.

Discover & Learn Human and Physical Geography — Volcanoes & Earthquakes

A Floating Crust

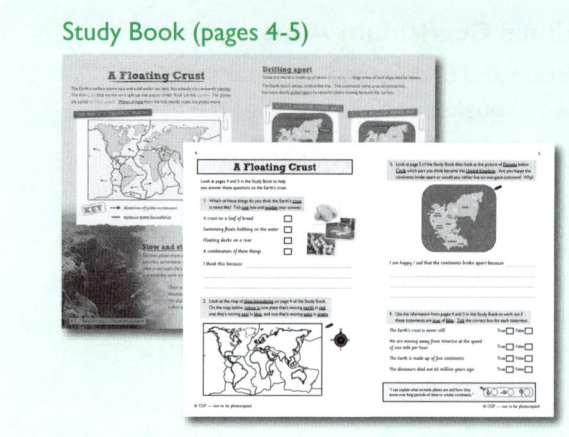

Study Book (pages 4-5)

Activity Book (pages 4-5)

National Curriculum Aims

Describe and understand volcanoes and earthquakes, including being able to:

- understand that the Earth's crust is split into tectonic plates that are able to move,
- understand that continents are shaped by plate tectonics.

Introduction

The meteorologist Alfred Wegener first proposed the idea of continental drift and the existence of a previous 'supercontinent' in 1912. His evidence came from such things as the layout of today's land masses, which seem to 'fit together', and geological records showing matching layers in rocks on different continents. Further evidence came from the fossils of similar extinct plants and animals being found on separate continents.

As a part of teaching this topic, it may be helpful to show pupils an online video that illustrates how the supercontinent of Pangaea broke apart and moved around to create the seven continents we have today: Asia, Africa, North America, South America, Europe, Australia/Oceania and Antarctica.

Answers to Activity Book Questions

1. Any ticked box with an appropriate explanation. E.g. *A crust on a loaf of bread. I think this because* the crust is thin and goes all the way around the Earth like the crust on a loaf of bread. / *Swimming floats bobbing on the water. I think this because* the Earth's tectonic plates float on the mantle like floats on water. / *Floating ducks on a river. I think this because* the Earth's crust is floating and moving about on the mantle like ducks on water.

2. Pupils should have shaded one plate moving north in red, one plate moving east in blue and one plate moving west in green, based on the direction of the arrows shown in the Study Book.

3. Pupils should have drawn a circle around the correct part of Eurasia — the small peninsular sticking up out of the western end of Eurasia (just below the larger area that would become Norway and Sweden). Any appropriate answer for why they might be happy or sad that the continents broke apart, e.g. *I am sad that the continents broke apart because* if they had stayed together I could visit other parts of the world more easily and cheaply, without having to get on an aeroplane or a boat.

4. True — False — False — True

Extra Activities

- Give each pair of pupils a map of the world and ask them to shade the sea in blue and the land in green. Then get them to label the seven continents, the UK and the Pacific and Atlantic Oceans.

- North America moves away from Europe at a rate of roughly 0.1 mm a day. In small groups, pupils could work out how long it takes for North America to move one millimetre, one centimetre, one metre and one kilometre away from Europe.

- Provide pupils with a printed map of the world. Get them to cut out the continents and stick them together to create a supercontinent similar to Pangaea. You could either ask them to recreate the shape of Pangaea shown on page 5 of the Study Book, or to create and name their own supercontinent.

Clashes and Collisions

Study Book (pages 6-7)

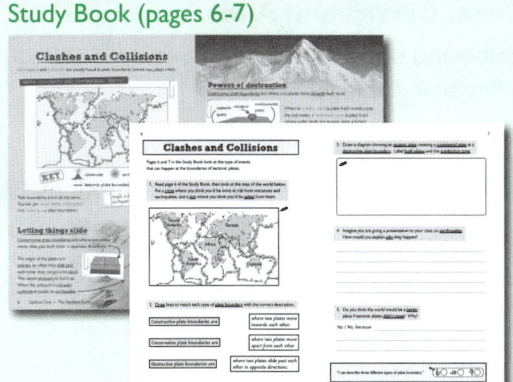

Activity Book (pages 6-7)

National Curriculum Aims

Describe and understand volcanoes and earthquakes, including being able to:

- describe the different types of plate boundaries,
- understand how different types of plate boundary give rise to earthquakes and volcanoes,
- use maps to investigate the physical features of plate boundaries around the world.

Introduction

It can be difficult to visualise the different types of plate interactions, so it may be useful to do the first 'Extra Activity' (below) with pupils as you discuss the different types of boundaries.

Tectonic plates moving towards one another at destructive plate boundaries can create fold mountains. The pressure of the plates colliding causes the crust to fold and push upwards. This led to the formation of the Alps and the Himalayas, and is still happening in these mountain ranges today. (You can illustrate this process for pupils by pushing two towels towards each other on a flat surface and showing how they fold when they meet.)

Answers to Activity Book Questions

1. Cross — any appropriate answer, i.e. a place on a plate boundary with many volcanoes and earthquakes.
   Tick — any appropriate answer, i.e. a place that is far away from any plate boundaries.

2. Constructive plate boundaries are — where two plates move apart from each other.
   Conservative plate boundaries are — where two plates slide past each other in opposite directions.
   Destructive plate boundaries are — where two plates move towards each other.

3. Pupils' drawings should show a portion of the oceanic plate under the continental plate. Pupils should have labelled the point where the oceanic plate is forced under the continental plate as the subduction zone.

4. E.g. earthquakes happen when one tectonic plate slides past or collides with another one. As the plates move, they can get stuck. This causes pressure to build up. When the plates move again, this pressure is released suddenly, causing an earthquake.

5. Either yes or no with any appropriate explanation. E.g. *Yes, because* we would have fewer natural disasters as there wouldn't be earthquakes or volcanic eruptions.

Extra Activities

- Use two gym mats to demonstrate the types of plate boundary. For conservative boundaries, put the mats side by side and move them slowly in opposite directions. For destructive boundaries, move them towards each other, pushing one under the other as they meet (the subduction zone). For constructive boundaries, move the mats apart. Ask pupils to discuss how the movement of the mats relates to real plate boundaries.

- One piece of new evidence that supported the theory of plate tectonics was that climbers who reached the summit of Mount Everest in the 1950s and 1960s brought back rock samples that contained marine fossils. Using the information in the introduction, discuss with pupils why finding marine fossils at the top of Mount Everest supports the theory of plate tectonics. Pupils could write a diary entry in the character of a scientist from the 1950s who studied the rock samples, describing their feelings when they discovered that the rocks contained marine fossils.

Discover & Learn Human and Physical Geography — Volcanoes & Earthquakes

Vesuvius

Study Book (pages 8-9)

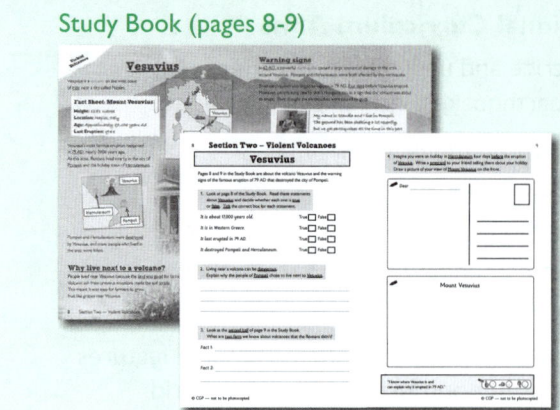

Activity Book (pages 8-9)

National Curriculum Aims

Describe and understand volcanoes, including being able to:

- describe a volcanic eruption,
- understand what causes volcanic eruptions,
- understand that there are signs that can help scientists predict a volcanic eruption.

Introduction

Mount Vesuvius is a volcano that lies on the western coast of Italy, very close to the city of Naples. It has erupted many times, although its eruptions have varied greatly in size and impact. It last erupted in 1944, and is currently going through a quiet period — though it will almost certainly erupt again at some point in the future.

The eruption of Vesuvius in 79 AD is one of the most famous volcanic eruptions in history. It buried the nearby settlements of Pompeii and Herculaneum under a thick layer of ash and volcanic debris.

Answers to Activity Book Questions

1. True — False — False — True

2. E.g. the people of Pompeii chose to live next to Vesuvius because the volcanic ash made the soil fertile / good for farming.

3. *Fact 1:* The movement of tectonic plates is what causes volcanoes.
   Fact 2: Some volcanoes, like Vesuvius, are on subduction zones (which can cause a build up of magma and then an eruption).

4. Any appropriate answer and drawing. Their postcard should reference the small earthquakes that occurred in the days leading up to the eruption. They could also talk about what makes Herculaneum a good holiday town, or mention that the land around Vesuvius was used for farming. Their picture should show a gently sloping mountain, like the pictures on page 8 of the Study Book. It might include buildings or fields to show the towns and farmland nearby. It should not show the volcano erupting.

Extra Activities

- Provide pupils with photos of other volcanoes in Italy, e.g. Etna, Stromboli and Vulcano, and ask them to plot these on a map. Ask pupils to make a poster about one of these volcanoes. They could include a timeline of its eruptions, the type of material it erupts and whether it is active or dormant. Pupils could also note how it has affected the people who live nearby, such as the benefits of tourism or damage from an eruption.

- Show pupils an online video of the newsreels showing the eruption of Vesuvius that occurred in 1944 and then ask them to write a newspaper article about the event. They could include eye-witness accounts, scientific explanations, or the eruption's relevance to the Second World War.

- Talk to pupils about the fact that Pompeii was a prosperous city, full of shops selling locally grown food and crafts. Ask pupils to imagine that they are shopping in the city on the day before the eruption. Divide the class into small groups to role play a shopkeeper and customers who are experiencing the minor earthquakes before the eruption of Vesuvius. What might they say to each other about the earthquakes? Would they be scared or worried? Would the shopkeeper be worried about their shop?

Timeline of an Eruption

Study Book (pages 10-11)

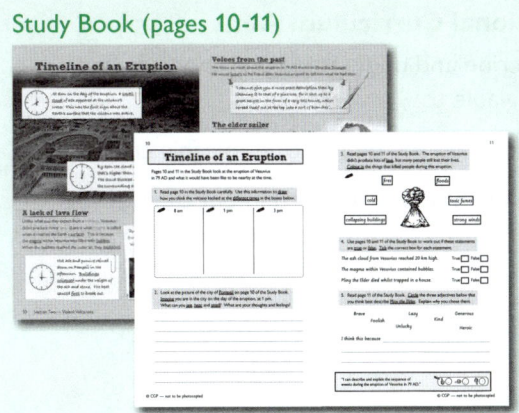

Activity Book (pages 10-11)

National Curriculum Aims

Describe and understand volcanoes, including being able to:

- describe the effects of a volcanic eruption,
- describe and understand the timeline of an eruption.

Introduction

This topic uses a contemporary eye-witness account to describe the sequence of events that took place during the eruption of Vesuvius in 79 AD.

After teaching pupils about Pliny the Younger's letters, you could discuss with them if they think these letters are a trustworthy source for information about the eruption, or if there is any reason why they may be limited in their reliability and/or usefulness.

Answers to Activity Book Questions

1. Any appropriate drawings relating to the descriptions on page 10 of the Study Book.
2. Any appropriate answer drawing on information from the Study Book and pupils' own ideas. E.g. they might talk about being able to see the huge ash cloud over Vesuvius, the sky going dark, smelling smoke or ash, and hearing the volcano rumbling. They might say they would feel scared, or wonder if the Gods were angry.
3. Pupils should have coloured: fires, collapsing buildings, toxic fumes.
4. True — True — False
5. Any three adjectives with an appropriate explanation.

Extra Activities

- During the eruption, houses collapsed under the weight of the ash and rock. Show some images of houses in Pompeii, pointing out the key features of their structures (e.g. pillars, central courtyard). Challenge pupils to make the strongest model building possible in groups of four, using only six sheets of A4 paper and sticky tape. Test how strong the models are by carefully pouring different amounts of sand over them.

- Read pupils the extract below from the letters of Pliny the Younger as he witnessed the eruption of Vesuvius:
"*The sea seemed to roll back upon itself, and to be driven from its banks by the convulsive motion of the earth; it is certain at least the shore was considerably enlarged, and several sea animals were left upon it. On the other side, a black and dreadful cloud, broken with rapid, zigzag flashes, revealed behind it variously shaped masses of flame...*"
Ask pupils to draw, paint or make models to illustrate this description.

- Read pupils the extract below from the letters of Pliny the Younger:
"*My mother now besought, urged, even commanded me to make my escape at any rate, which, as I was young, I might easily do; as for herself, she said, her age and corpulency rendered all attempts of that sort impossible; however, she would willingly meet death if she could have the satisfaction of seeing that she was not the occasion of mine.*"
Ask pupils to create a short drama about this story and perform it for their classmates.

The Eruption Continues

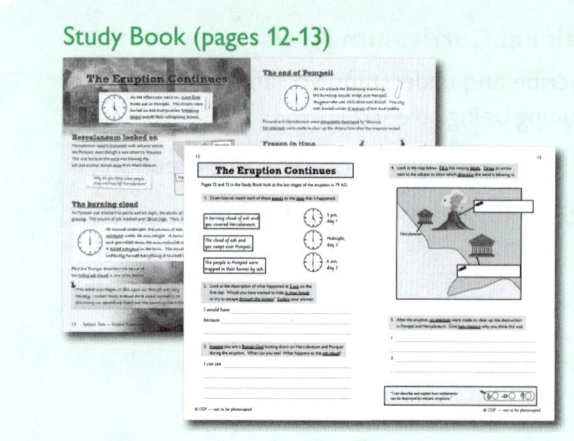

Study Book (pages 12-13)

Activity Book (pages 12-13)

National Curriculum Aims

Describe and understand volcanoes, including being able to:

- describe the effects of a volcanic eruption,
- understand that a settlement can be destroyed by a volcanic eruption.

Introduction

The eruption of Vesuvius in 79 AD buried Pompeii in over 5 metres of ash and rubble and Herculaneum in 20 metres, preserving them exactly as they were at the time of the eruption. Because of this, modern excavations of them have given archaeologists lots of information about what life was like in Italy during the Roman Empire.

Answers to Activity Book Questions

1. A burning cloud of ash and gas covered Herculaneum. — Midnight, day 2
   The cloud of ash and gas swept over Pompeii. — 6 am, day 2
   The people in Pompeii were trapped in their homes by ash. — 5 pm, day 1

2. Any appropriate answer. E.g. *I would have* tried to escape through the streets *because* I wouldn't want to get trapped in my house. / *I would have* stayed in my house *because* I'd want to avoid falling pumice stones.

3. Any appropriate answer relating to the ash cloud falling over Herculaneum and then Pompeii.

4. *From top to bottom:* Vesuvius, Pompeii.
   The arrow to show the direction of the wind should point south-east (from Vesuvius towards Pompeii).

5. Any appropriate answers. E.g. everyone was killed so there was no one left to clear up the destruction / there was too much damage.

Extra Activities

- Explain to pupils that some of what we know about life in Pompeii comes from signs and graffiti found on the walls of buildings around the city. Challenge them to use a beginners' Latin book or the internet to learn how to say some basic phrases in Latin. For example, challenge them to learn how to say "Hello. My name is _____ and I am _____ years old." Then ask them to introduce themselves to a friend in Latin.

- Show the class aerial photographs of the ruins of Pompeii and Herculaneum and discuss what the buildings were used for (e.g. temples, theatres, shops, baths and houses). You could show the pupils artefacts that were found in some of the buildings to provide clues. Pupils could label what the different buildings were for on maps of the two settlements, or make their own models of objects that were found in the buildings.

- Pupils could create a drama or dance about the destruction of Pompeii. They could begin by doing everyday activities such as shopping or talking with friends. Then the volcano erupts, the ash cloud rises into the sky and pumice stones start to fall. Ask the pupils how they would feel about the events happening around them and how they would respond. The performance could be set to music and may be in slow motion so pupils can focus on expressing their emotions through facial expressions and body movements.

Types of Volcano

Study Book (pages 14-15)

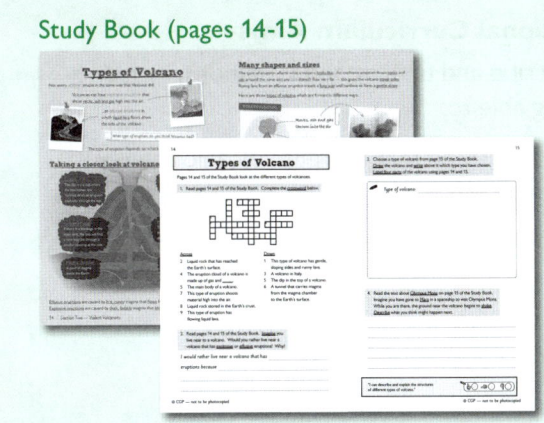

Activity Book (pages 14-15)

National Curriculum Aims

Describe and understand volcanoes, including being able to:

- describe the key features of a volcano,
- describe different types of volcano,
- understand how volcanoes are formed.

Introduction

Before teaching this topic, you could ask pupils whether they've seen any volcanoes in real life or in documentaries. If they have, get them to describe what shape the volcanoes were and/or what happened when how they erupted.

As there is a lot of new vocabulary on these pages in the Study Book, it might be useful to familiarise pupils with the key words before starting the Activity Book pages.

Answers to Activity Book Questions

1. Across: 2 Lava, 4 Ash, 5 Cone, 7 Explosive, 8 Magma, 9 Effusive.
   Down: 1 Shield, 3 Vesuvius, 5 Crater, 6 Vent.

2. Any appropriate answer. E.g. *I would rather live near a volcano that has* explosive *eruptions because* I wouldn't have to worry about the lava flow.

3. Any appropriate drawing of a stratovolcano, shield volcano or a cinder cone with four correct labels.

4. Any appropriate account. E.g. Lava starts to flow down the sides of Olympus Mons from its crater. The lava is hot and runny because Olympus Mons is a shield volcano. The lava is flowing quickly towards us so we get back in the spaceship and set off again, just as the lava reaches the place where we landed.

Extra Activities

- Pumice is formed when lava filled with bubbles of volcanic gas is cooled quickly. Put pupils in small groups and give them a pumice stone to observe closely with a magnifying glass. Ask them to sketch the stone and make predictions about: a) how much it weighs, b) whether it will float on water, and c) how hard it is (can it be scratched by other hard objects?). Then get them to test their predictions.

- In small groups, pupils could make a papier-mâché volcano. To do this, take an empty 500 ml plastic bottle and put it in the centre of a square of thick cardboard. Scrunch up some newspaper and place it around the bottle in a cone shape, using masking tape to keep the newspaper in place. Cover the whole structure with papier-mâché, then leave it to dry. Paint with acrylic paint and then cover with a clear, waterproof varnish. To make the lava, put four teaspoons of bicarbonate of soda in the bottle and add 150 ml of warm water mixed with red food colouring. Add vinegar to make the volcano erupt. Make sure to leave plenty of space around the volcano for the 'lava' to flow into. (Video demonstrations of this process can be found online.)

- Get pupils to research volcanic activity on other planets and moons in our solar system (examples with either past or current volcanic activity include Mars, Venus, Io, Triton and Enceladus). Split pupils into groups and give each group one planet or moon to investigate. They could finish by giving short presentations to the rest of the class, describing the planet or moon they were investigating, including whether its volcanoes are active or extinct and what is/was released during the eruptions.

Volcanoes in the Deep

Study Book (pages 16-17)

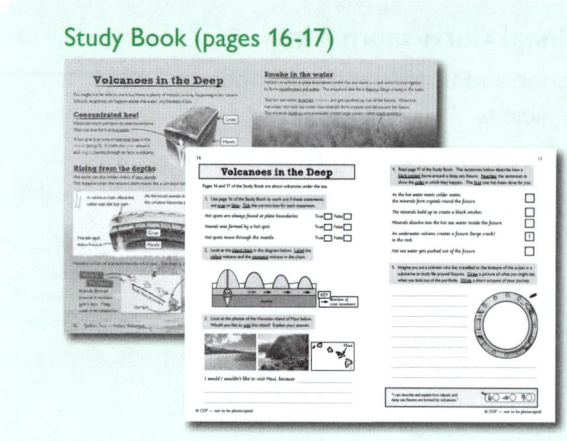

Activity Book (pages 16-17)

National Curriculum Aims

Describe and understand volcanoes, including being able to:

- understand that there is volcanic activity underwater,
- describe how volcanoes can create islands.

Introduction

Pupils may find it hard to imagine volcanoes erupting underwater, but in fact there are far more volcanoes underwater than on land. The world's largest known volcano, Tamu Massif, is under the Pacific Ocean. It is a huge, extinct shield volcano that covers an area of over 100,000 square miles (making it larger than the UK).

Answers to Activity Book Questions

1. False — True — False

2. Pupils should have labelled the volcano above the hot spot (on the left) as the youngest volcano, and the volcano furthest from it (on the right) as the oldest volcano.

3. Pupils can answer either way, provided they justify their answer sensibly. They could draw on information from the Study Book and the images, e.g. *I would like to visit Maui because it has nice beaches, and I like to swim and surf. / I wouldn't like to visit Maui because it has active volcanoes, and I would be scared of eruptions.*

4. Order: 4, 5, 2, *1*, 3.

5. Any appropriate drawing and account. Pupils should draw on information from the Study Book. E.g. pupils may mention seeing fissures in the rocks, black smokers and deep sea creatures like giant tube worms.

Extra Activities

- Pupils could do some research about Hawaii Island. Split the class into small groups and give each group a subject to research such as tourism, climate, wealth and poverty, wildlife, plastic pollution, other forms of pollution, earthquakes or volcanic activity. Afterwards, pupils could put the facts and images they've found into columns of pros and cons to visiting Hawaii Island. Using these columns, the class could debate whether or not they would like to travel to Hawaii Island before taking a final vote.

- Show pupils images of black smokers and some of the creatures that live around them (e.g. tube worms, vent clams, hydrothermal octopuses, squat lobsters). Get pupils to create a wall frieze of an undersea fissure, complete with black smokers and the creatures around it. Pupils could draw or paint individual items to be cut out and stuck on to a prepared background.

- Ask pupils to think of as many ways as they can that new islands can be formed. Examples include volcanic activity below the sea (e.g. Hawaii), land being separated from a continent due to changes in sea level (e.g. Great Britain), coral islands (e.g. the atolls of the Maldives) and man-made islands (e.g. Palm Jumeirah in Dubai or Kansai Airport's island in Japan). Show pupils maps or aerial photographs of some of these examples. Can they see any clues that could tell them how these islands formed? (For example, volcanic islands are often found in chains of small islands. Islands that were separated from continents are usually close to a large land mass. Coral atolls are ring shaped. Man-made islands often have unusual or very regular shapes.)

San Francisco

Study Book (pages 18-19)

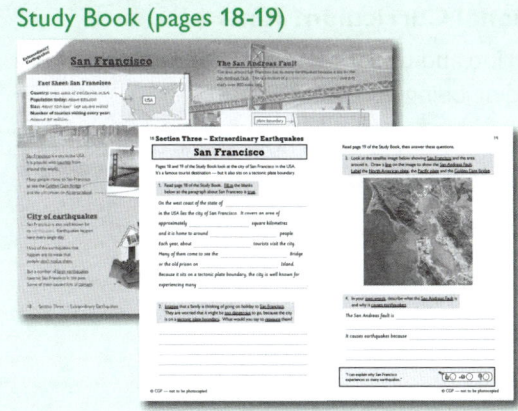

Activity Book (pages 18-19)

National Curriculum Aims

Describe and understand earthquakes, including being able to:

- explain why earthquakes occur where they do,
- understand that some settlements are located near to plate boundaries.

Introduction

San Francisco is a hilly city in Northern California. It began as a small Mexican settlement called Yerba Buena in an area called Alta California, which became part of the USA in 1846. Its population expanded rapidly in the gold rush of 1849. Today it is famous for being the home of many tech companies — 'Silicon Valley' is nearby.

Before teaching this topic, it may be useful to recap with the pupils how earthquakes happen.

Answers to Activity Book Questions

1. *On the west coast of the state of* California *in the USA lies the city of San Francisco. It covers an area of approximately* 120 *square kilometres and it is home to around* 880,000 *people. Each year, about* 25 million *tourists visit the city. Many of them come to see the* Golden Gate *Bridge or the old prison on* Alcatraz *Island. Because it sits on a tectonic plate boundary, the city is well known for experiencing many* earthquakes.

2. Any appropriate answer. E.g. although there are lots of earthquakes, most of them are very small / the family could have a plan in case of an earthquake.

3. Approximately correct placement of the San Andreas Fault, and correct labelling of the North American plate, the Pacific plate and the Golden Gate Bridge.

4. E.g. *The San Andreas Fault is* part of the plate boundary between the Pacific and North American plates. *It causes earthquakes because* tension is built up by the two plates moving past each other, which is then released as an earthquake.

Extra Activities

- Give pupils a list of between five and ten major earthquakes that have happened in California in the past 150 years (this information should be available online). Give each pupil a map of California and ask them to plot the earthquakes. They could add a numbered key to help identify the earthquakes.

- Ask pupils to make their own timeline of major earthquakes in the San Francisco Bay Area (they should be able to find information about this online). They could add details about each earthquake to their timeline, such as the magnitudes and how much damage each one caused. They could also note when the strongest earthquake was and whether there were any quiet periods (where no major earthquakes occurred).

- Have a class discussion about the dangers and benefits of living near a plate boundary. Show pupils a map of the world with plate boundaries and some major cities marked on, noting any that lie near a plate boundary. Ask them to think about why so many people live in places near to plate boundaries. (Page 27 of the Study Book mentions some benefits to living near a plate boundary. Pupils may mention other ideas, such as having family or a job there. If they are struggling, get them to think about why their own family lives where they do.)

The 1906 Earthquake

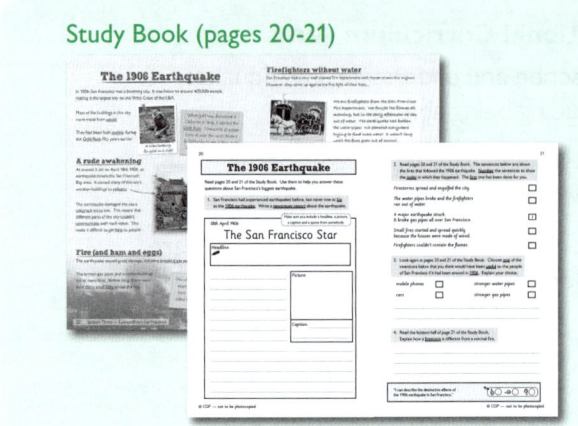

Study Book (pages 20-21)

Activity Book (pages 20-21)

National Curriculum Aims

Describe and understand earthquakes, including being able to:

- describe and explain how an earthquake can cause damage,
- understand how people may respond to an earthquake.

Introduction

There was just one foreshock to the 1906 earthquake, only 25 seconds before the earthquake struck. This meant the people in San Francisco had barely any warning, and not really any time to do anything to prepare for it. To introduce pupils to this topic, ask them what they think it might feel like to experience a large earthquake like the one that happened in San Francisco in 1906. They might talk about things like seeing buildings falling down, cracks appearing in the ground and the ground shaking very strongly.

Answers to Activity Book Questions

1. Any appropriate report about the 1906 San Francisco earthquake. Answers could include an eye-witness account, details about damage to buildings, and details about the fires that followed.
2. Correct order: 5, 3, *1*, 2, 4.
3. Any appropriate answer. E.g. mobile phones, because if everyone had phones they would have been able to contact each other and get help to people who needed it.
4. E.g. a firestorm is a fire so big that it sucks in the air around it, creating a strong wind around the fire. The winds around the firestorm can also make it spread faster than a normal fire.

Extra Activities

- Due to the gold rush in California, San Francisco was a 'boomtown'. Get the pupils to discuss how they think the discovery of gold around San Francisco affected the area in terms of human activity and the impact it had on the natural environment. Ask them to consider the differences in living conditions between the rich business owners and the less well-off people, such as the miners. Pupils could use the internet to look for pictures and maps of San Francisco before and after the gold rush, as well as after the earthquake. They could then create a presentation or a wall display with these images and their thoughts about San Francisco in 1906.

- The damage to San Francisco's communication systems during the 1906 earthquake made it hard for people to relay information or get help. Get pupils to design a board game where the aim is for a messenger to deliver an important message about the earthquake damage to the other side of the city. Different squares on the board can pose different problems for the messenger, e.g. 'a burning building is blocking your path, miss a turn', 'a burst gas pipe slows you down, go back three spaces', or 'a local resident shows you a shortcut, go forward six spaces'.

- Pupils could be put into small groups and plan short dramas to be performed one after the other. Divide the story of the earthquake into sections and give one section to each group. The sections could be: gold-panning in the river, building a wooden house, the earthquake striking, cooking breakfast and fire breaking out, firefighters battling the flames and running out of water.

Discover & Learn Human and Physical Geography — Volcanoes & Earthquakes

After the Earthquake

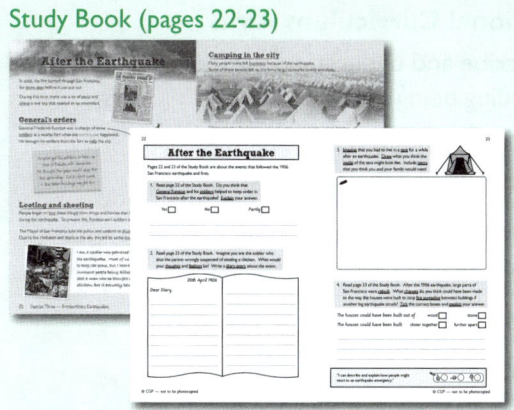

Study Book (pages 22-23)

Activity Book (pages 22-23)

National Curriculum Aims

Describe and understand earthquakes, including being able to:

- describe and explain how an earthquake can cause damage,
- describe and explain how people may respond to an earthquake emergency.

Introduction

About 80% of the city of San Francisco was destroyed in the earthquake and subsequent fires of 1906. Emergency action had to be taken to save lives and restore order. This included reinforcing the law as people began to loot, but it also meant finding food and shelter for people who had lost their homes.

Many people lived in tents in parks and had to stand in long queues for food. Others were made to cook on the streets instead of inside their houses in order to prevent more fires from breaking out.

Answers to Activity Book Questions

1. Any appropriate explanation. E.g. *partly* because the soldiers patrolling the streets may have stopped some people from looting. But some innocent people were shot.

2. Any appropriate diary entry that includes the thoughts and feelings of the soldier. Answers could include feeling guilty / sad / regretful / shocked. Thoughts could include that the soldier was following orders / the Mayor gave bad orders / the soldier genuinely believed the man was looting.

3. Any appropriate drawing that includes essentials for living, such as bedding, food, water and medicine.

4. Any appropriate answer. E.g. wider spaces were left between houses to prevent fires travelling / materials that are less flammable were used (i.e. not wood).

Extra Activities

- Pupils could investigate charities that help people who have been displaced by natural disasters, e.g. the Red Cross, ShelterBox and Tearfund. (Other charities can be found on the Disasters Emergency Committee website.) Discuss ways in which pupils could help to raise money for disaster relief charities, such as holding an exhibition of their work or having a bake sale.

- Find images of groups of people involved in the 1906 San Francisco earthquake. Ask pupils to look at the images and discuss in groups what they think the people were thinking and feeling. This could also be a way to start pupils writing a short story or diary entry about the event.

- Pupils could create a mixed media piece of art to show the difference between the San Francisco skyline before the earthquake and fire in 1906 and how it looks today. They could begin by making a sunset background, created with either paint or pastels. Then the pupils could use black paper to make the modern skyline and put this in front of the background. Finally, pupils could make the wooden houses from 1906 out of sticks, wood shavings or brown paper strips and place these in front of the modern skyline.

Discover & Learn Human and Physical Geography — Volcanoes & Earthquakes

Earthquake Information

Study Book (pages 24-25)

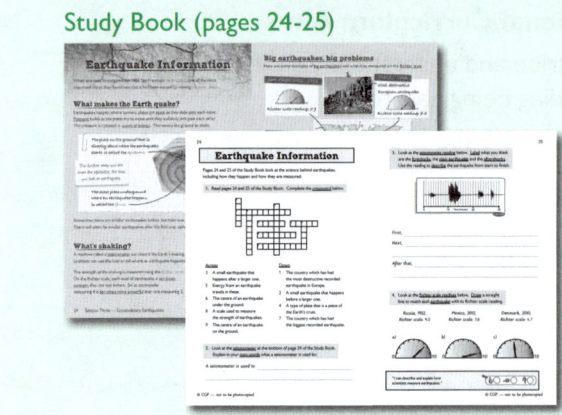

Activity Book (pages 24-25)

National Curriculum Aims

Describe and understand earthquakes, including being able to:

- describe how the strength of an earthquake can be measured,
- explain how scientists can predict where earthquakes may occur,
- understand that it's hard to predict when earthquakes may occur.

Introduction

Scientists are quite good at predicting where earthquakes are likely to occur, but it isn't currently possible to predict exactly when they will happen or how big they will be. Small earthquakes happen frequently, but big earthquakes that cause a lot of damage are relatively rare. The Richter scale is one commonly used scale for classifying the strength of earthquakes, although other scales, like the Mercalli scale, can be used too.

Answers to Activity Book Questions

1. Across: 3 Aftershock, 5 Waves, 6 Focus, 8 Richter, 9 Epicentre.
   Down: 1 Italy, 2 Foreshock, 4 Tectonic, 7 Chile.

2. E.g. *A seismometer is used to* measure how strong an earthquake is.

3. The small waves between 2 and 5 minutes are the foreshocks, the biggest waves between 10 and 15 minutes represent the main earthquake, and the two sets of smaller waves from 22-24 minutes and 27-28 minutes are the aftershocks.
   Any appropriate description. E.g. *First,* there is one foreshock. *Next,* the earthquake happens at 10 minutes and lasts 5 minutes. *After that,* there are two main aftershocks. The first lasts two minutes, the second lasts one minute.

4. a) Mexico, 2012, Richter scale: 7.6, b) Russia, 1952, Richter scale: 9.0, c) Denmark, 2010, Richter scale: 4.7

Extra Activities

- Search online for a list of the ten largest earthquakes ever recorded. Give this information to pupils and ask them to work in pairs to mark the earthquakes on a map of the world. They could also add the tectonic plate boundaries to the map and write a short paragraph to explain the link between the size of an earthquake and its proximity to a plate boundary.

- Pupils could research another scale apart from the Richter scale that is used to measure the size of earthquakes (e.g. the Mercalli scale, which measures earthquakes on the basis of how strongly they are felt and how much damage they cause). Pupils could display their findings on a poster.

- Small groups of pupils could make their own seismometers. To do this, take a shoe box and cut a large rectangle in the top of the lid, leaving a 2 cm border around the edge. Link several small elastic bands together to make two chains, then stretch the chains over the shoe box diagonally (so that they make a taut X-shape over the top). Next, make a horizontal slit on each of the two ends of the box 1 cm up from the bottom and position a long strip of paper through these slits. Place a marker pen in the centre of the elastic band cross, so that the pen sits upright at the centre of the box and just touches the paper. As you pull the paper strip through the slits at the end sides of the box, the pen will record any vibrations on the paper. (Full video demonstrations of how to make a seismometer like this can be found online.)

Troubled Earth

Study Book (pages 26-27)

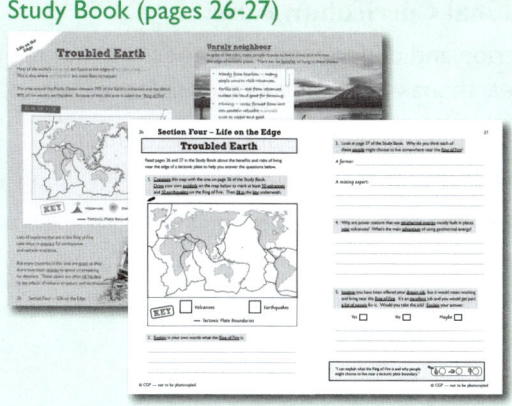

Activity Book (pages 26-27)

National Curriculum Aims

Describe and understand volcanoes and earthquakes, including being able to:

- understand that volcanic eruptions and earthquakes are most likely to happen near plate boundaries,
- describe some of the benefits to living in places with lots of tectonic activity,
- explain how tectonic activity affects land use and the location of settlements.

Introduction

The Ring of Fire is a 40,000 km arc that stretches around the boundary of a large part of the Pacific plate. There are over 450 active and dormant volcanoes along the Ring of Fire. But many millions of people live in this region, putting them at risk from the impacts of volcanic eruptions and earthquakes.

Answers to Activity Book Questions

1. The map should show any 10 volcanoes and any 10 earthquakes from the map on page 26 of the Study Book. The symbols for the key should match the symbols they have drawn on their map.

2. E.g. the Ring of Fire is an area that runs all around the edge of the Pacific Ocean/plate, which is the site of a large number of volcanoes and earthquakes.

3. *A farmer:* e.g. volcanic soil is very fertile, which would help farmers to grow crops.
   A mining expert: e.g. there are valuable minerals in the rocks near volcanoes, so there are likely to be lots of mines and jobs for mining experts.

4. E.g. in places near volcanoes, the ground is very hot not far below the surface, so it's easy to get at the heat. The main advantage of using geothermal energy is that it is renewable / it will never run out.

5. Any appropriate answer with sensible justification. E.g. Yes, because if it's my dream job with good pay, it will be worth the risk of experiencing earthquakes and volcanic eruptions.

Extra Activities

- Since volcanic soils tend to be fertile, many countries in the Ring of Fire are major food producing nations. As a class, pick one of the countries in the Ring of Fire that has a large agricultural industry (e.g. the Philippines, Indonesia or Chile). Bring in examples of foods grown there for pupils to try, e.g. peaches, pears and corn from Chile, coconut, mango and pineapple from the Philippines, and cocoa, tea and cinnamon from Indonesia. Pupils could choose one crop from the country (not limited to those brought in to class) and make a small poster about it. The posters could form a class display about agriculture in that country.

- Pupils could make a table of information about the countries that are affected by the Ring of Fire. They could use the map on page 26 of the Study Book and an atlas or the internet to gather key facts about the countries, including area, population size, life expectancy, CO_2 emissions and average income. You could display Google Earth™ on a screen and have pupils take it in turns to rotate the globe in order to find the names of the countries nearest to the Ring of Fire.

Keeping Safe – Prediction

Study Book (pages 28-29)

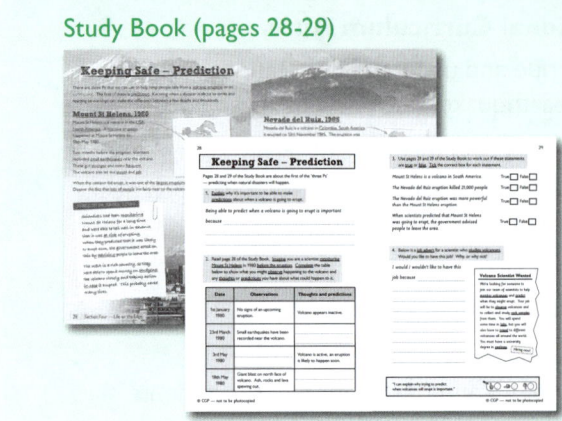

Activity Book (pages 28-29)

National Curriculum Aims

Describe and understand volcanoes and earthquakes, including being able to:

- describe some of the signs that scientists can use to predict a volcanic eruption,
- understand how predicting a natural disaster can help to save lives.

Introduction

This topic looks at what signs indicate that a volcano might be about to erupt and how scientists can use this information to warn people that an eruption may be coming. These warnings can save many lives.

Despite being a comparatively small volcanic event, the eruption of Nevado del Ruiz in 1985 was the worst natural disaster in Colombia's history, due in large part to a failure in emergency protocols and evacuations. This failure prompted the creation of the Volcano Disaster Assistance Program (VDAP). This organisation helps to monitor the world's active volcanoes and predict eruptions, reducing the risk that a country's wealth might impact its ability to respond to the threat of a volcanic eruption.

Answers to Activity Book Questions

1. E.g. *Being able to predict when a volcano is going to erupt is important because* it can give the government time to warn people who live near the volcano to leave before it erupts.

2. E.g. *1st January 1980, No signs of an upcoming eruption. Volcano appears inactive.*
   23rd March 1980, Small earthquakes have been recorded near the volcano. Volcano might be active, an eruption is possible.
   3rd May 1980, Steam and ash rising from volcano. Volcano is active, an eruption is likely to happen soon.
   18th May 1980, Giant blast on north face of volcano. Ash, rocks and lava spewing out. Volcano is erupting.

3. False — True — False — True

4. Any appropriate answer that refers to details in the advert and/or draws on information from the Study Book.

Extra Activities

- As a follow-up to looking at Mount St Helens and Nevado del Ruiz, pupils could discuss the differences and similarities between the geography of Colombia and the USA. They could find the two countries in an atlas and investigate their different geographical features such as forests, mountain ranges and land usage. They could discuss how easily the area around each volcano could have been evacuated. E.g. jungles and mountains would make it more difficult, and densely populated areas would be harder to evacuate.

- Ask pupils to investigate the techniques volcanologists (scientists who study volcanoes) use to do their jobs, such as measuring vibrations in the earth and recording gas levels around volcanoes. Videos can be found online explaining these methods. Pupils could write a diary entry of a day in the life of a volcanologist.

- Give pupils a list of four important signs that a volcano might erupt soon: gas leaking from the volcano, bulges in the side of the volcano, lots of small earthquakes near it, and steam coming from the top. Ask pupils to make a musical composition to depict these signs using a range of instruments such as whistles (for the gas leaks and steam), drums (for the bulges), and maracas (for the small earthquakes).

Keeping Safe – Prevention

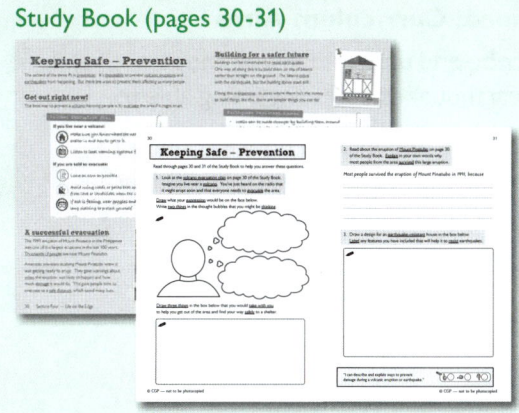

Study Book (pages 30-31)

Activity Book (pages 30-31)

National Curriculum Aims

Describe and understand volcanoes and earthquakes, including being able to:

- understand some of the steps that can be taken to prevent harm to people and property during a volcanic eruption or earthquake.

Introduction

Although preventing volcanic eruptions is currently impossible, NASA is trying to find ways to do just that. Underneath Yellowstone National Park in the USA lies a supervolcano that could erupt in the future. If it did, its effects would be catastrophic. Theories of how to prevent it from erupting focus on the idea of cooling the magma chamber down, so that it never becomes hot enough to trigger an eruption.

Until we can come up with guaranteed ways of preventing natural disasters though, we need to put in place measures that will help to keep people safe when they do happen.

Answers to Activity Book Questions

1. Any appropriate answers. E.g. the expression of the face could show shock, fear or worry. The thoughts could be about it being important to leave the area quickly, remembering the evacuation plan, wanting to make sure your family and friends are safe, worrying about personal safety or damage to your house. Drawings of things to take with you when evacuating could include e.g. goggles, long clothing, torches, maps, mobile phones, radios, water/food.

2. E.g. *Most people survived the eruption of Mount Pinatubo in 1991, because* scientists had been studying the volcano and they knew it was going to erupt soon. This gave people enough time to evacuate and get to a safe distance.

3. Any appropriate labelled drawing. Pupils could draw on information from page 31 of the Study Book, e.g. they might include beams between the floor and the ground to allow movement, a thin metal roof, a strong frame inside the walls of the house, or straw-packed walls.

Extra Activities

- Get pupils to imagine that a volcano is predicted to erupt in their local area and that they are reporters covering the story on a local news channel. They could create a bulletin to warn the local population of the possible eruption and advise them of the steps they should take to protect themselves.

- Nepal and the Philippines have very different landscapes: Nepal is a mountainous, landlocked country, but the Philippines is an island country. Pupils could look at aerial photographs of both countries and discuss the different challenges of evacuating people from an area in each of the two countries during a natural disaster.

- Get pupils to look at the pictures of earthquake-proof buildings on page 31 of the Study Book. Put pupils into groups and give them some materials to make simple models of earthquake-resistant buildings. They could use e.g. lolly sticks, modelling clay, card, newspaper and glue. They could then test their structures by carefully placing weights on top and/or shaking or banging on the table to represent an earthquake.

Keeping Safe – Preparation

Study Book (pages 32-33)

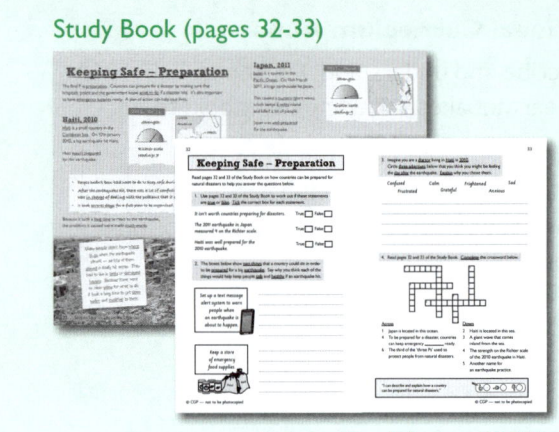

Activity Book (pages 32-33)

National Curriculum Aims

Describe and understand volcanoes and earthquakes, including being able to:

- describe and explain how a country can prepare for natural disasters, such as volcanic eruptions or earthquakes.

Introduction

How well-off a country is can be a factor in determining how well prepared it is to deal with natural disasters, such as earthquakes. Pages 32 and 33 in the Study Book look at the effects of a large earthquake in Japan (which has the 3rd largest economy in the world, based on GDP) and Haiti (which has the 143rd largest economy in the world, based on GDP).

Answers to Activity Book Questions

1. False — True — False.

2. *Set up a text message alert system to warn people when an earthquake is about to happen*: e.g. it gives people time to find somewhere safe to go during the earthquake.
   Keep a store of emergency food supplies: e.g. people who have had to leave their homes and have no food can be given something to eat.

3. Any appropriate answer. E.g. frightened, frustrated, grateful. Frightened because the earthquake was dangerous and you don't know what's going on, frustrated because you are able to help others but you can't get to them, grateful that you survived / can do something to help.

4. Across: 1 Pacific, 4 Supplies, 6 Preparation
   Down: 2 Caribbean, 3 Tsunami, 4 Seven, 5 Drill.

Extra Activities

- Discuss with pupils what roles rescue workers might perform after an earthquake (e.g. searching for people who are trapped, freeing them, giving medical treatment). You could go on to teach or remind pupils of some basic first aid (e.g. putting someone in the recovery position or dressing a wound).

- Show pupils some video footage of rescue workers helping people who are trapped in their houses after an earthquake. Discuss the difficulties of rescuing people after an earthquake, including the need for silence (so they can listen for trapped people making sounds), digging carefully (so as not to cause injury to people being rescued) and the risk of aftershocks.

- Put pupils into small groups and ask them to talk about what an earthquake drill might consist of and how they think it would differ from a fire drill (an earthquake drill usually involves getting under a table and putting your hands over your head until the 'quake' is over, then quietly going outside and lining up, as for a fire drill). Then show them footage of pupils in Japan doing an earthquake drill (this can be found online). The class could then have a go at recreating an earthquake drill in their classroom.

All Calm in the UK 1

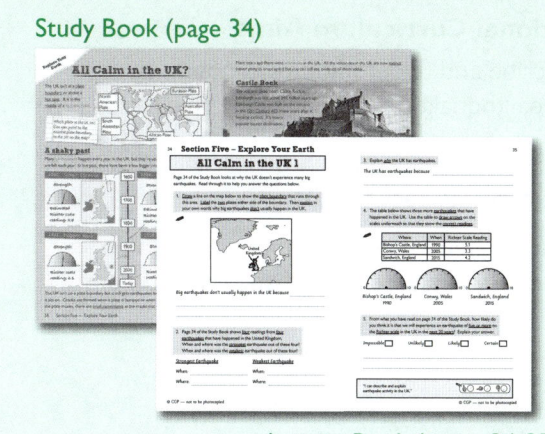

Study Book (page 34)

Activity Book (pages 34-35)

National Curriculum Aims

Describe and understand volcanoes and earthquakes, including being able to:

- describe where the United Kingdom lies in relation to plate boundaries,
- understand how earthquakes can occur away from a plate boundary.

Introduction

This topic gives pupils an opportunity to see where the UK sits in relation to plate boundaries. Before teaching this topic, it may be helpful to remind pupils that most earthquakes and volcanic eruptions happen at plate boundaries.

Answers to Activity Book Questions

1. The plate boundary line should look similar to the one on page 34 of the Study Book, i.e. running down to the east of Greenland, through Iceland and down through the Atlantic Ocean to the west of the UK.
   E.g. *Big earthquakes don't usually happen in the UK because* the UK is not on a plate boundary.

2. Strongest Earthquake — *When:* 1931, *Where:* Dogger Bank.
   Weakest Earthquake — *When:* 1884, *Where:* Colchester.

3. E.g. *I think the earthquakes were caused by* cracks in the plate that the UK sits on. As the plate moved, the cracks moved a little bit too.

4. The arrows should point to the correct numbers on the scale according to the table.

5. Either 'Unlikely' or 'Likely' should be ticked and an appropriate explanation given. E.g. *Unlikely* because in the UK, earthquakes over 5 on the Richter scale are not very common, so it's unlikely that we will have another in the next 20 years. / *Likely* because there are many earthquakes a year in the UK, and we have had earthquakes over 5 on the Richter scale before, so it's possible there will be another one soon.

Extra Activities

- Show pupils a newspaper report or some video footage covering an earthquake in the UK and point out any eye-witness accounts. Get them to write their own newspaper reports about a UK earthquake, including some eye-witness accounts. You could ask each pupil to create a character who was affected by the earthquake and interview one another as eye-witnesses for their reports. They could make up a story about where they were when the earthquake struck, e.g. they were in bed and were woken by the shaking.

- Get the pupils to look again at the map at the top of page 34 of the Study Book. Using the internet, a globe or an atlas, ask the pupils to make a list of ten countries which are touching a plate boundary on the map, and another list of ten countries which, like the UK, are not. Pupils could imagine that a plate boundary ran close to where they lived and discuss in small groups what the pros and cons would be moving to another part of the world if this were so. They could write up their ideas individually and then share them with the class.

- Search online for a list of earthquakes that have happened in the UK since 2014. Give pupils a printed map of the UK and get them to mark on the map where the earthquakes happened. Ask them to work out where the nearest of these earthquakes was to them. Did they feel it? If so, can they describe what it was like? If not, can they imagine what it would have been like?

All Calm in the UK 2

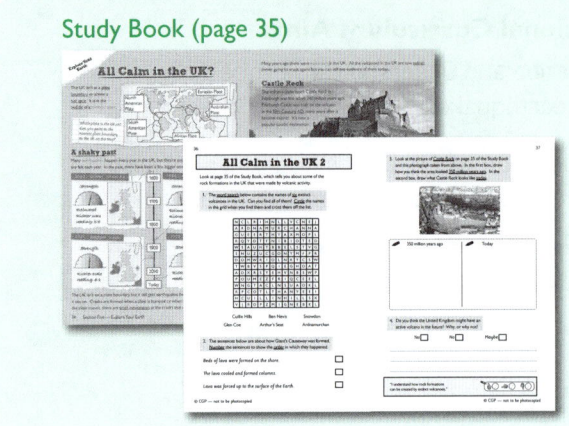

Study Book (page 35)

Activity Book (pages 36-37)

National Curriculum Aims

Describe and understand volcanoes and earthquakes, including being able to:

- describe and explain how volcanic activity can create rock formations,
- understand that the amount of tectonic activity in an area can change over time.

Introduction

Page 35 of the Study Book explores two rock formations in the UK that were shaped by volcanic activity millions of years ago. Further examples include the Inner Hebrides, which was once a chain of volcanoes, and Ben Nevis, the remains of a huge, ancient volcano that collapsed in on itself with astonishing force.

Answers to Activity Book Questions

1. Here are the locations of the six extinct volcanoes in the grid:

2. Beds of lava were formed on the shore. — 2
   The lava cooled and formed columns. — 3
   Lava was forced up to the surface of the Earth. — 1

3. Any appropriate drawings showing the volcano:
   - active at *350 million years ago*,
   - extinct at *100 million years ago*,
   - as pictured in the photograph provided with the question, i.e. with Edinburgh Castle on top of it, for *Today*.

4. Any answer is acceptable as long as it is explained sensibly. E.g. *Maybe* because the edges of plates can change / the UK used to be on a plate boundary, so it could be on one again.

Extra Activities

- Many of the columns that make up Giant's Causeway are roughly hexagonal. Show the pupils pictures of Giant's Causeway from a number of angles so they understand the shapes of the columns. Then get pupils to work in groups to make hexagonal prisms of different heights from card and join them together to make their own Giant's Causeway. They could also paint a sea and sky background to put the columns in context.

- Provide each pupil with a printed blank map of the UK. Ask them to use an atlas (or the internet) to find the locations of the six extinct volcanoes given in the word search on page 36 of the Activity Book (Cuillin Hills, Ben Nevis, Snowdon, Glen Coe, Arthur's Seat and Ardnamurchan). Ask pupils to plot the locations of each of these extinct volcanoes on their map.

- Read pupils the legend about Finn McCool and Giant's Causeway. Pupils could use this as inspiration to write their own legend about the creation of Castle Rock, or one of the other natural phenomena described in the Study Book (such as The Ring of Fire around the Pacific Ocean or Mount Vesuvius in Italy).

The Solar System

Study Book (pages 2-3)

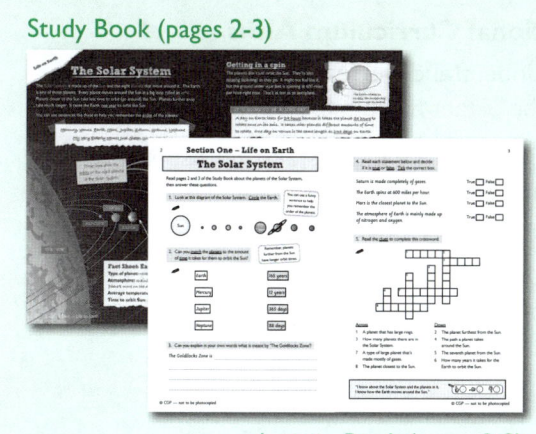

Activity Book (pages 2-3)

National Curriculum Aims

- Understand the Solar System and the Earth's position in it.
- Understand that the Earth and other planets in the Solar System orbit the Sun.
- Understand that the Earth is rotating on its axis.

Introduction

This topic allows pupils to see Earth in its wider context as part of the Solar System before they begin learning about the planet itself. Although pupils may assume that they are sitting still, the Earth is spinning on its axis at 600 miles per hour while flying through space at 67,000 miles per hour around the Sun. On top of that, the Solar System itself is moving around the Milky Way at 448,000 miles per hour.

Once pupils have read pages 2 and 3 of the Study Book, discuss what they think it would be like to go into space. You could show pupils videos of astronauts living aboard the International Space Station and how they go about their daily lives. Discuss how different it would be from life on Earth. Would pupils like to live there? Why or why not?

Answers to Activity Book Questions

1. Pupils should have circled the third planet from the Sun.
2. Earth — 365 days, Mercury — 88 days, Jupiter — 12 years, Neptune — 165 years
3. E.g. *The Goldilocks Zone is* an area in a solar system where planets can support life because it's not too hot or too cold / it's not too close or too far from the Sun. Earth is in the Goldilocks Zone in our Solar System.
4. False — True — False — True
5. Across: 1 Saturn, 3 Eight, 7 Gas Giant, 8 Mercury
   Down: 2 Neptune, 4 Orbit, 5 Uranus, 6 One

Extra Activities

- Ask pupils to work in pairs and create their own mnemonic for remembering the names of the planets in the Solar System, e.g. 'My Very Excited Mother Just Served Us Noodles'.
- Ask pupils to make a fact book about one of the planets in the Solar System. Ask them to research and record information such as: planet type, atmosphere, average temperature, time to orbit the Sun, the length of a day, any moons the planet may have and how the planet has been observed/recorded.
- Pupils can use Styrofoam balls to represent the Sun and the planets in the Solar System. Colour or paint each ball to look like the planet (or Sun) it represents. The planets can be joined together with cord or string to make a visual representation of the planets and their position relative to the Sun. Higher level pupils could research the distances of each planet from the Sun and make the distances to scale.

Discover & Learn Human and Physical Geography — Living Planet

Why Earth?

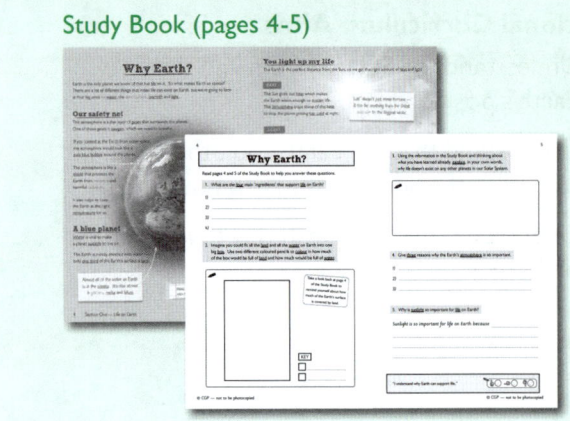

Study Book (pages 4-5)

Activity Book (pages 4-5)

National Curriculum Aims
- Understand the features of Earth that make the planet habitable.
- Describe and understand key aspects of human geography and natural resources, including food and water.
- Understand that day and night are a result of the Earth's rotation.

Introduction

This topic introduces pupils to the broad reasons why life exists on Earth. The combination of the atmosphere, water and Earth's distance from the Sun make it ideal for supporting life. There is also a magnetic field that surrounds the Earth, created by the molten iron at the centre of the planet. This field protects us from electromagnetic radiation from the Sun. Without any of these factors, complex life such as plants, animals and humans could not have evolved. The chances of life occurring as we know it were unimaginably small, which may explain why we haven't discovered life on any other planets yet.

Once pupils have read pages 4 and 5 of the Study Book, ask them if they think there are other planets in the universe with life on them. What might life on other planets look like? Would they look anything like humans?

Answers to Activity Book Questions

1. Water, atmosphere, warmth, light
2. Pupils should have coloured one third of the box to represent land, and two thirds to represent water.
3. E.g. Life doesn't exist on any other planets in the Solar System because they are either too hot or too cold / there isn't enough light / the atmosphere doesn't have the right gases / there isn't enough water.
4. E.g. Any three of the following: It contains the oxygen we need to breathe / It protects the Earth from meteors. / It protects the Earth from harmful radiation. / It keeps the Earth at the right temperature.
5. E.g. *Sunlight is so important for life on Earth because* plants need it to grow, and if we didn't have plants, humans and other animals would be unable to survive.

Extra Activities

- Pupils could scatter seeds on damp cotton wool and then place them in different conditions (i.e. one in light and cold, one in light and warm, one in dark and cold and one in dark and warm) to measure how well they grow depending on their conditions. Pupils could also make up their own investigations, e.g. sowing seeds in one box with a small hole to let a pin-prick of light in and in another box with a slightly larger hole, to investigate how little sunlight is necessary to allow a healthy plant to grow.

- Ask pupils to consider the idea of living on Mars. Pupils could design their own 'Martian Village'. What would it look like? What would they need to survive? How would they grow food? How would they breathe? How would they make water? They could then build their village using modelling clay or cardboard.

- Ask pupils to research what the different gases are that make up the Earth's atmosphere. They could present their findings as a diagram or pie chart.

What Time Is It?

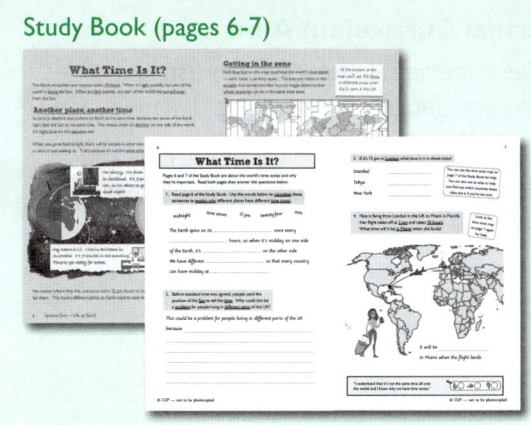

Study Book (pages 6-7)

Activity Book (pages 6-7)

National Curriculum Aims
- Identify the position and significance of the Prime Meridian Line and Greenwich Mean Time.
- Understand the significance of time zones.
- Understand that day and night are a result of the Earth's rotation.

Introduction

Before Greenwich Mean Time was established, settlements across the UK used sundials to determine the time. Slight differences and inaccuracies across the UK meant that no two places used exactly the same time. When people began travelling by rail, these slight differences made it difficult to make an accurate timetable. In 1840, Railway time was introduced. Most railway companies used GMT as Railway time, but it wasn't until 1880 that GMT was adopted as standard time in the UK.

This topic introduces pupils to time zones and standardised time and explains why we need them. Once pupils have read pages 6 and 7 of the Study Book, ask them if they have ever been abroad on holiday. If so, can they find where they went on the world map on page 7? Is it in a different time zone? Do they remember noticing the time difference? (They might have had to put their watches forward or remember having jet lag.)

Answers to Activity Book Questions

1. *The Earth spins on its* axis *once every* twenty-four *hours, so when it's midday on one side of the Earth, it's* midnight *on the other side. We have different* time zones *so that every country can have midday at* 12 pm.

2. Pupils should recognise that different places in UK will have been using slightly different times. This was a problem when people began to travel more, particularly by train, and the time wasn't the same everywhere.

3. Istanbul: 3 pm (+ 3 hours),   Tokyo: 9 pm (+ 9 hours),   New York: 7 am (− 5 hours)

4. *It will be* 7 pm *in Miami when the flight lands.*

Extra Activities

- Get pupils to make clocks to display in class that show what the time is in different cities or countries when it's 12 pm in London/the UK.
- Get pupils to cut out an image of the world's time zones (as on page 7 of the Study Book) and attempt to wrap it around a small ball. Alternatively, they could look at a globe with time zones marked on it. They could then see how the zones apply to the spherical nature of the Earth.
- Using a football (the Sun), a tennis ball (the Earth) and a torch (the light from the Sun), ask pupils to explain to their peers how day and night occur. Higher level pupils could attempt to explain how the Earth's tilt results in the seasons.
- Ask pupils to imagine that they are trying to set up a group video call which must include them, as well as someone living in New York, and someone in Beijing. Ask them to work out what time they should organise the call to give them the best chance of everyone being awake. If they complete this, get them to write their own questions of a similar type and use them to test each other's understanding.

Discover & Learn Human and Physical Geography — Living Planet

More Than Weather

Study Book (pages 8-9)

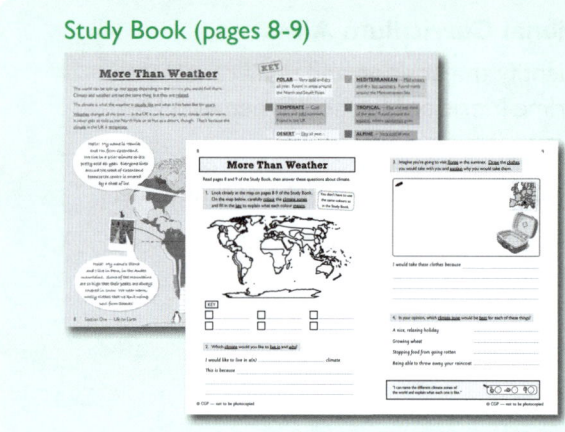

Activity Book (pages 8-9)

National Curriculum Aims
- Describe and understand key aspects of physical geography, including climate zones.
- Understand geographical similarities and differences between the UK and other countries.

Introduction

Earth can be divided into a number of different climate zones, roughly determined by how far they are from the equator — with tropical climates being the closest to the equator and polar climates being the farthest away. The climate zones also have an effect on what plants and animals can live in a particular place. Pupils will investigate this in more detail in Section Three.

This topic gives pupils the opportunity to consider how other countries differ from the UK and what effects the climate might have on the way of life of people living in other countries.

Answers to Activity Book Questions

1. Map and key should be shaded in so that each area of the map matches the correct climate zone in the key.

2. Any appropriate answer. Pupils should draw on information from the Study Book. E.g. *I would like to live in a tropical climate. This is because the weather is hot for most of the year and I'd like to visit the rainforest.*

3. Pupils' drawings should show clothing suitable for dry, hot weather, e.g. shorts, t-shirts, dresses, skirts, sunglasses, hats, swimsuits. Pupils' explanation should show that they understand that Rome has a Mediterranean climate, and as a result will be hot and dry in the summer.

4. E.g. *A nice, relaxing holiday* — Mediterranean / tropical / temperate,
Growing wheat — temperate / Mediterranean, *Stopping food from going rotten* — Polar / Alpine,
Being able to throw away your raincoat — Desert / Polar

Extra Activities

- Get pupils to make a card game by writing the names of the different climate zones on one set of cards and descriptions of the climates on another set. The aim of the game is to match the climate zone to its description. If each pupil has a set of climate zone cards, one pupil could read out a description and the first pupil to put down the matching climate zone card wins. The game could also be played in the style of a game of 'snap'.

- Ask pupils to create a 'Fauna Map' of the world. Pupils can research what animals live in various climate zones, both on land and in the sea. Pupils could then stick pictures of the animals to the appropriate areas on a large world map.

- Ask pupils to consider what it might be like to live in a different country with a different climate. Have any pupils travelled abroad and experienced different climate types? Ask pupils to imagine they have a pen-pal in a country with a different climate to the UK. Get them to write a letter asking their pen-pal questions about the what the climate is like where they live and how it affects their life. E.g. 'What types of clothes do you wear?', 'What do you like about the weather where you live?', 'Are there any special precautions you have to take because of the weather where you live?'

Ocean Life

Study Book (pages 10-11)

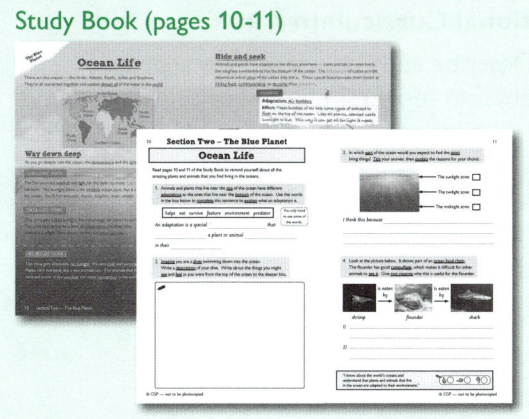

Activity Book (pages 10-11)

National Curriculum Aims
- Be able to describe the location of the world's oceans and identify their key physical characteristics.
- Describe and understand key aspects of climate zones and biomes.

Introduction

The oceans cover about 70% of the Earth's surface. This topic introduces pupils to the idea that the organisms living in different zones (depths) of the oceans have special adaptations that help them to survive there.

Charles Darwin published his theory of evolution in 1859. It states says that individuals with features that help them survive in their environment are likely to live longer and have more offspring. They pass on these features to their offspring. So over time, more individuals will end up with features that make them well-adapted to their environment. Adaptation and evolution are covered in the Year 6 content of the Key Stage 2 science syllabus.

Start this topic by asking pupils what challenges they think ocean creatures might face in their habitat. Examples of things to contend with include light levels, temperature, finding food and escaping predators.

Answers to Activity Book Questions

1. *An adaptation is a special* feature *that* helps *a plant or animal* survive *in their* environment.
2. Any appropriate answer. Pupils should draw on information from the Study Book. Answer should include that it would get darker and colder the deeper they went down. Pupils may also include descriptions of the different types of plants and animals they would see in the upper, middle and lower parts of the ocean.
3. Pupils should have ticked: The sunlight zone. E.g. *I think this because* the sunlight zone gets more warmth and light from the Sun than the other zones, so more species of animals and plants can live in it.
4. E.g. to make it easier for the flounder to sneak up on its prey without being seen / to make it easier for the flounder to hide from predators .

Extra Activities

- Ask pupils to imagine they are on a diving trip going down into the twilight zone of the ocean. On the trip, they discover a sea creature that has never been seen before. Get them to think about what the creature might be like (e.g. what size it is, what colour it is, what it might eat). Ask pupils to name their creature and to draw a picture of it, labelling any features it has that make it well adapted to its environment.

- Get pupils to pick a sea creature to research. Ask them to find out what adaptations their chosen creature has that help it survive in its environment. Examples of creatures that they could research include: dolphins, moon jellyfish, puffer fish, common squid, sea turtles, seahorses, blue whales and limpets.

- Divide the class into pairs and give them pictures of some sea life — for example, cards showing seaweed (or phytoplankton), krill, clam, squid, herring, shark and seal. Challenge pupils to choose some of the pictures and arrange them into a food chain. Ask them to give reasons why they have arrived at their answer. Pupils could then combine their food chains with those created by other pairs to make a food web.

Discover & Learn Human and Physical Geography — Living Planet

Oceans and the Climate

Study Book (pages 12-13)

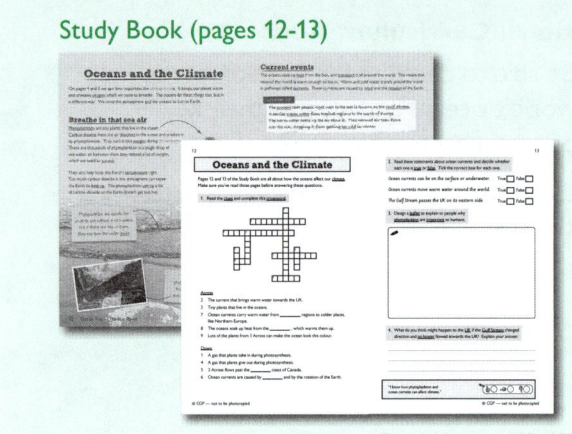

Activity Book (pages 12-13)

National Curriculum Aims
- Describe and understand key aspects of physical geography, including climate zones and biomes.

Introduction

The oceans have a huge effect on the world's climate. Oceans absorb heat energy from the sun. Ocean currents then redistribute this heat energy around the world, carrying warmer water towards the polar regions and colder water toward the equator. The oceans also take in a lot of carbon dioxide, partly by simply dissolving it and partly due to marine plants taking it in to use for photosynthesis. This keeps large amounts of carbon dioxide (a greenhouse gas) out of the atmosphere, so it helps to combat global warming.

Answers to Activity Book Questions

1. Across: 2 Gulf Stream, 3 phytoplankton, 7 tropical, 8 Sun, 9 green
   Down: 1 carbon dioxide, 4 oxygen, 5 east, 6 wind

2. True — True — False

3. Any appropriate answer. Pupils should draw on information from the Study Book. Points they could include are that phytoplankton take up carbon dioxide, release oxygen and help to maintain the Earth's temperature.

4. E.g. the UK might start to have very cold winters, because without the Gulf Stream there wouldn't be as much warm air blowing over the UK.

Extra Activities

- Search online for a video clip telling the story of the 28 000 rubber ducks that fell overboard from a ship going from Hong Kong to the USA in 1992. Show the video to the class, then give each pupil a printed copy of a world map with the site of the duck spill marked. Show them a list of places and dates that some of the ducks washed up — you could include southern Alaska in 1992, northern Japan in 1995, eastern Australia in 1996, south-east Canada in 2001, western Scotland in 2003 and north-west France in 2007. Get them to plot these on their maps. Ask the class what they think this information could tell us about our oceans — answers could discuss where ocean currents flow, how fast they travel and how long plastic waste lasts in the oceans.

- Many types of phytoplankton are single celled organisms with interesting shapes. Search online for images of phytoplankton (diatoms are a good type to search for). Show pupils some of the images you have found and point out how many different shapes there are. Ask pupils to create a picture of one of the phytoplankton, using any appropriate materials (felt-tip pens, paint, felt, glue, string etc.). The pictures can be used to make a class display.

Discover & Learn Human and Physical Geography — Living Planet

Mountains

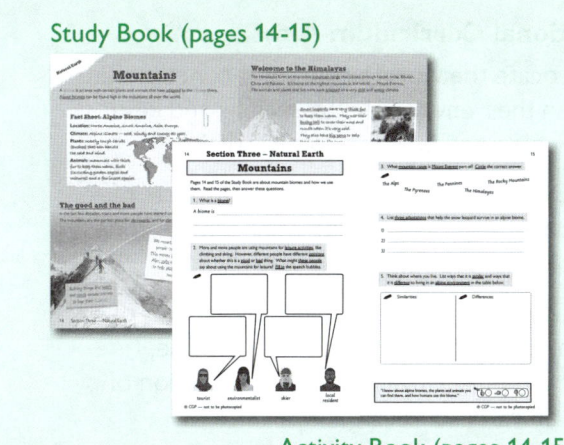

Study Book (pages 14-15)

Activity Book (pages 14-15)

National Curriculum Aims

- Locate the world's countries, concentrating on their environmental regions and key human and physical characteristics.
- Describe and understand key aspects of physical geography, including climate zones, biomes and mountains.
- Describe and understand key aspects of human geography, including land use and economic activity.

Introduction

Alpine biomes are usually found below the snow line on mountains, but above 3000 m in altitude. They are often relatively inhospitable, with high winds and freezing temperatures. However, these areas are still home to plants, animals and humans.

This topic introduces pupils in detail to the first of ten land-based biomes in this section, focusing on the plants and animals that live there, their adaptations to the environment and human interactions with the biome.

Answers to Activity Book Questions

1. *A biome is* an area that contains particular plants and animals that have adapted to the climate there.

2. Pupils may answer in any way, as long as they give sensible reasons for their answers. E.g. a tourist may be against building hotels and ski slopes because it spoils the view, or be in favour because it gives her a place to stay. The environmentalist may be against mountains being used for leisure because of the damage to landscape, plants and animals caused by tourists and building tourist attractions. The skier may be glad of the ski slopes and hotels because it provides a tourist destination for her. The local resident may dislike the number of tourists visiting and damaging the landscape, or be in favour of the money it brings to the area.

3. Pupils should have circled: The Himalayas.

4. E.g. thick fur (to keep warm) / bushy tail (to keep warm) / big paws (to walk on snow).

5. Any appropriate answer. Pupils should draw on information from the Study Book. E.g. *Similarities*: In the UK and in alpine biomes there are tourist attractions / plants, animals and people live there / there are mammals, insects, birds, amphibians and reptiles. *Differences*: In alpine biomes, the weather is snowy and windy all year / there are large wild animals like snow leopards / the land is very mountainous / there aren't many tall trees.

Extra Activities

- Ask pupils to design a new creature that would be perfectly suited to living in the mountains. What is its coat like? E.g. furry or smooth. What type of feet does it have? E.g. hooves, paws or talons. What does it eat? E.g. animals, plants or both. Does it have any special qualities? Where does it find shelter?

- Ask pupils to imagine they're a climber about to tackle Mount Everest. What equipment would they need? Pupils should choose five things they think they should take with them and explain why they chose each one.

- Ask pupils to design a tourist information board for visitors to a mountain area, to tell them what plants and animals they might see. Pupils could use the information in the Study Book to focus on the Himalayas, or you could provide them with another mountain range (e.g. Alps, Andes or Rocky Mountains) to research.

Discover & Learn Human and Physical Geography — Living Planet

Tropical Rainforests

Study Book (pages 16-17)

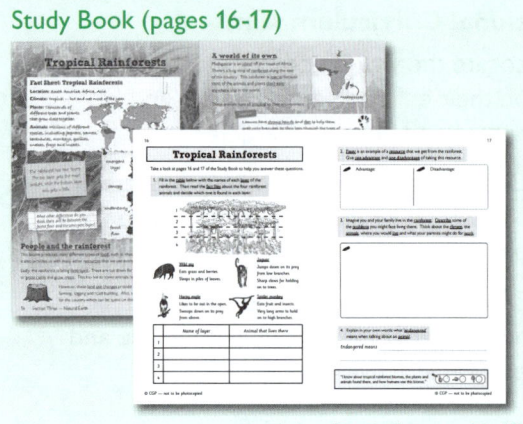

Activity Book (pages 16-17)

National Curriculum Aims

- Locate the world's countries, concentrating on their environmental regions and key human and physical characteristics.

- Describe and understand key aspects of physical geography, including climate zones, biomes and vegetation belts.

- Describe and understand key aspects of human geography, including land use, economic activity and the distribution of natural resources.

Introduction

This topic explores the tropical rainforest biome which is home to more species of animals and plants than any other biome on Earth.

The rainforest can be divided into four distinct layers. The emergent layer is comprised of the few trees that compete for sunlight by growing taller than most. The second highest of these layers, the canopy, is home to most of the animal species in the rainforest, including monkeys, birds and sloths. This layer is so dense that it blocks most of the light from reaching the layers below. The understorey receives enough light to support small plants like shrubs. But the forest floor gets almost none, and for that reason doesn't support much plant life. Instead it's covered in leaf litter and other plant debris and is home to many insect species.

Answers to Activity Book Questions

1. 1) Emergent layer, Harpy eagle, 2) Canopy, Spider monkey 3) Understorey, Jaguar
   4) Forest floor, Wild pig.

2. E.g. *Advantage*: Paper is a very useful resource that we use every day. / Making paper creates jobs.
   Disadvantage: Making paper means you have to cut down large areas of the rainforest. / Cutting down trees destroys habitats for rainforest animals.

3. Any appropriate answer. Pupils should draw on information from the Study Book. E.g. They would have to be careful of big predators like jaguars. It would rain a lot so they would need a shelter to keep the rain out. There aren't many jobs available, so their parents might have to hunt, fish or collect resources to sell.

4. *Endangered means* at risk of becoming extinct.

Extra Activities

- Ask pupils to imagine that they are a rainforest explorer. What can they see, hear, taste, touch and smell? What emotions are they feeling? Pupils could summarise their ideas in the form of a diary extract describing a day spent exploring the rainforest.

- Provide pupils with data about the temperature and rainfall in a tropical rainforest and in the UK. Ask them to compare the two and decide which climate they'd most like to live in.

- Ask pupils to choose one of the animals mentioned in the Study Book — jaguar, parrot, tarantula, monkey, gorilla, snake, frog, chameleon, lemur or mantella. Pupils can make a fact file about their chosen animal, focussing on: where it lives in the rainforest, what it eats, what its predators are, how it is adapted to survive in a rainforest and whether or not it's endangered. As a follow-up activity, pupils could compare fact files and decide which animal is most endangered by looking at figures such as how many of each animal is left in the wild and how much of their habitats remain.

Discover & Learn Human and Physical Geography — Living Planet

Woodlands

Study Book (pages 18-19)

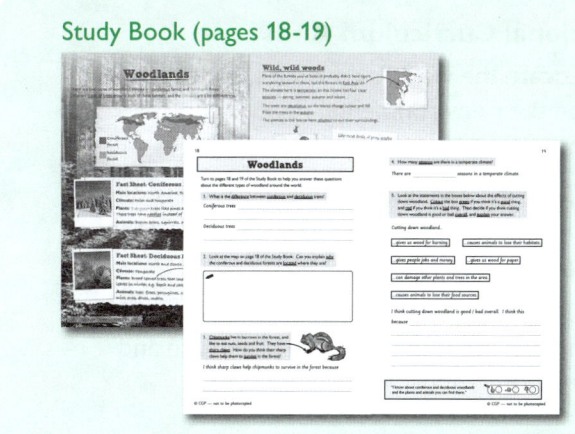

Activity Book (pages 18-19)

National Curriculum Aims
- Locate the world's countries, concentrating on their environmental regions and key human and physical characteristics.
- Describe and understand key aspects of physical geography, including climate zones, biomes and vegetation belts.
- Describe and understand key aspects of human geography, including land use, economic activity and the distribution of natural resources.

Introduction

This topic introduces pupils to two types of forest biome — coniferous forests and deciduous forests. The majority of the natural woodland across the UK is deciduous forest. Most coniferous woodland in the UK has been artificially created, e.g. for timber production.

Once pupils have read pages 18 and 19 of the Study Book, ask them if they have ever been in a forest in the UK. What was it like? Were all the trees deciduous or were there coniferous trees too? Did they see any animals?

Answers to Activity Book Questions

1. E.g. *Coniferous trees* have needles instead of leaves. They keep their needles all year round.
   Deciduous trees have broader leaves and lose their leaves in autumn.

2. E.g. coniferous forests are found in colder and drier climates because they're adapted to survive colder temperatures. Deciduous forests are found in slightly warmer, wetter climates.

3. E.g. *I think sharp claws help chipmunks to survive in woodlands because* they help them to climb trees to look for food / they help them to dig their burrows.

4. *There are* four *seasons in a temperate climate.*

5. Pupils should have coloured green: ...gives us wood for burning. / ...gives people jobs and money. / ...gives us wood for paper.
   Pupils should have coloured red: ...causes animals to lose their habitats. / ...can damage other plants and trees in the area. / ...causes animals to lose their food sources.
   Pupils may answer either way, as long as they give sensible reasons for their answer. E.g. *I think cutting down woodland is* good *overall. I think this because* it gives us fuel, paper and wood which we need for lots of things.

Extra Activities

- Show pupils pictures of deciduous woodlands in the autumn (or use those on pages 18 and 19 of the Study Book as examples). Ask them to create their own autumnal forest scene using paint or colouring pencils.

- As a class, investigate deciduous and coniferous trees in more detail. Pupils could either go to a local park or bring collected leaves into class. Ask them to identify as many types of tree as they can from the leaves. Pupils can also sort the leaves (or needles) into deciduous and coniferous species.

- Split pupils into pairs and ask them to imagine they're going camping together in the woods for the night. They have food, water and a tent, but nothing else. Provide pupils with the following list of items and ask them to pick three items they would take with them: torch, matches, sleeping bag, rucksack, game, sleeping mat, radio, walkie-talkie, sweets, comic. Pupils can discuss as a class why they chose those items.

Discover & Learn Human and Physical Geography — Living Planet

Hot, Cold and In-Between

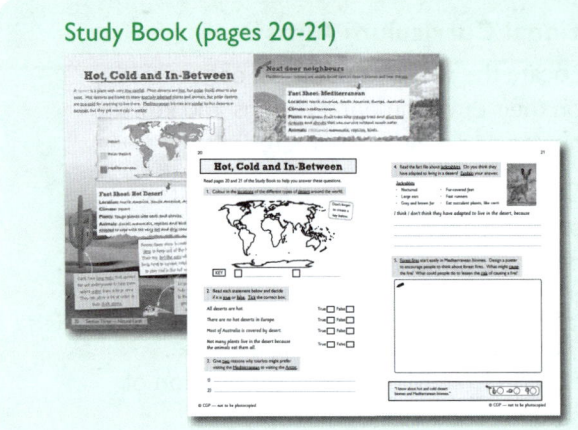

Study Book (pages 20-21)

Activity Book (pages 20-21)

National Curriculum Aims

- Locate the world's countries, concentrating on their environmental regions and key human and physical characteristics.

- Describe and understand key aspects of physical geography, including climate zones, biomes and vegetation belts.

- Describe and understand key aspects of human geography, including land use and economic activity.

Introduction

A desert is defined as an area without much vegetation or rainfall. However, deserts are not limited to places of extreme heat. Despite being covered with snow, many places in the Arctic and Antarctic receive so little annual precipitation that they are also considered to be deserts.

Unlike their closest relative, the tundra biome, polar deserts are so cold and dry that they cannot support any plant or animal life whatsoever. This topic describes hot deserts and Mediterranean biomes in more detail.

Answers to Activity Book Questions

1. Pupils' drawings should match the map on page 20 of the Study Book showing the locations of desert and polar desert biomes (but not the Mediterranean biomes).

2. False — True — True — False

3. E.g. the weather is warmer / there are beaches / you can swim in the sea / there are more holiday resorts.

4. Pupils should identify that jackrabbits have adapted to a desert climate. They should draw on information in the Study Book to connect some of the jackrabbit's features with living in a hot, dry climate. E.g. it's nocturnal to avoid the hottest part of the day and has large ears to keep cool. / Its grey and brown fur helps it to stay camouflaged in the desert. / It eats succulent plants full of water to stay hydrated. / Its fur-covered feet protect its skin from the hot sand. / It's a fast runner to escape predators in the open desert.

5. Many forest fires are caused by people being careless with fire. Pupils could draw, e.g. people using matches or campfires irresponsibly (or responsibly), or trees being destroyed by fires.

Extra Activities

- Provide pupils with data on the average monthly rainfall in the Mediterranean, a polar desert and a hot desert. Pupils could use the data to make graphs and compare them.

- Give pupils a list of desert plant adaptations and how they benefit the plant. Without telling them which adaptation matches which benefit, ask pupils to write the adaptations on one set of cards and the benefits on another. Pupils can then match the adaptation card to its corresponding benefit.
  E.g. Waxy leaves — seal water inside the leaves so it doesn't evaporate.
  Sharp spines — shade the surface of the plant and stop animals from eating it.
  Shallow roots — let plants absorb water when there's only a small amount of rain.
  Water stores — allow plants to survive during long dry spells.

- Ask pupils to draw their own desert animal species. What adaptations would it have? What colour would it be? Pupils can label their drawings to point out the animal's adaptations.

Grasslands

Study Book (pages 22-23)

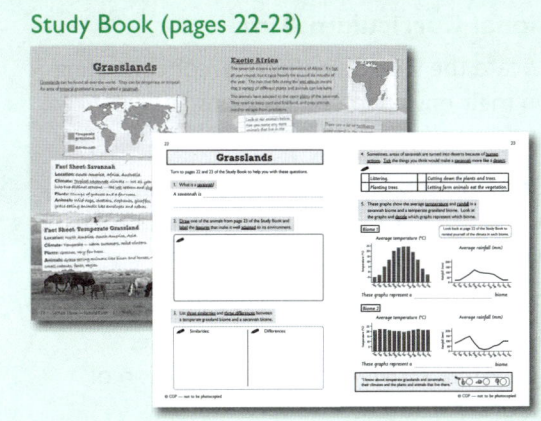

Activity Book (pages 22-23)

National Curriculum Aims

- Locate the world's countries, concentrating on their environmental regions and key human and physical characteristics.

- Describe and understand key aspects of physical geography, including climate zones, biomes and vegetation belts.

- Describe and understand key aspects of human geography, including land use and economic activity.

Introduction

This topic compares two types of grassland — temperate and tropical grassland. Temperate grasslands are also known as plains or prairies (particularly in North America) and tropical grasslands are often called savannahs. In spite of their name, some temperate grasslands can range in temperature from −18 °C to 32 °C between winter and summer. In contrast, savannahs generally stay between 20 °C and 30 °C all year.

Answers to Activity Book Questions

1. *A savannah is* an area of tropical grassland.

2. Any appropriate drawing. Pupils should draw on information from the Study Book for their labels.

3. Any appropriate answer. Pupils should draw on information from the Study Book. E.g.
Similarities: Both biomes have a lot of grasses. / Neither biome has a lot of trees. / Both biomes have grass-eating animals. / Both biomes can be found in South America.
Differences: Savannahs are hot all year, temperate grasslands are cooler in winter. / Savannahs have some trees, temperate grasslands have very few. / Savannahs have a wet season and a dry season, temperate grasslands don't. / Savannahs are found in Africa and Australia, but temperate grasslands aren't. / Temperate grasslands are found in North America and Asia, but savannahs aren't.

4. Pupils should have ticked: Cutting down the plants and trees and Letting farm animals eat the vegetation.

5. Biome 1: *These graphs represent a* temperate grassland *biome.*
Biome 2: *These graphs represent a* savannah *biome.*
Pupils should identify that Biome 1 has cooler winters and warmer summers and no drastic variations in rainfall. They should also identify that Biome 2 has a higher temperature all year round, and has a distinct rainy season.

Extra Activities

- Ask pupils to design a leaflet telling people about the issue of illegal hunting (poaching) and how it affects animals like rhinos, lions and elephants. How might they help to stop this? Pupils could make campaign posters for a wildlife charity, or create a radio or TV advertisement to draw attention to the issue.

- Get pupils to create their own 'Grassland Display'. Make a large background of blue and green/brown to represent the sky and savannah. Pupils could add animals to the display that they have drawn or painted. Once finished, discuss the ways in which the savannah is different from the countryside of the UK.

- Ask pupils to choose an animal that lives on the savannah and create a fact file on that animal, including: its full name (and its Latin name), what it eats, what adaptations it has, its predators and how endangered it is.

Discover & Learn Human and Physical Geography — Living Planet

A Frozen Place

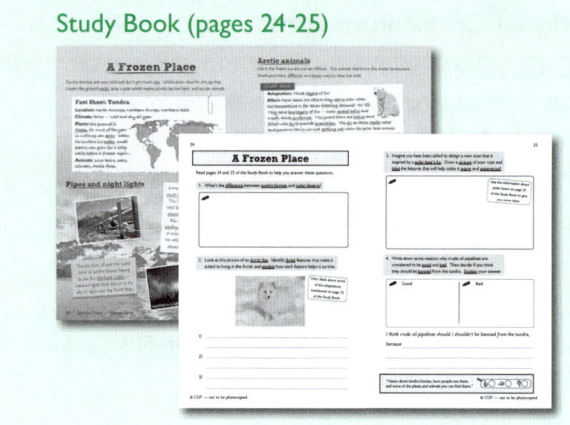

Study Book (pages 24-25)

Activity Book (pages 24-25)

National Curriculum Aims
- Locate the world's countries, concentrating on their environmental regions and key human and physical characteristics.
- Describe and understand key aspects of physical geography, including climate zones, biomes and vegetation belts.
- Describe and understand key aspects of human geography, including land use, economic activity and the distribution of natural resources.

Introduction

This topic introduces pupils to tundra biomes. While they may appear similar to a polar desert, tundras go through cycles of freezing and thawing which allows them to support plant and animal life.

Answers to Activity Book Questions

1. E.g. The ground in a polar desert is frozen all year. The ground in the tundra remains frozen for most, but not all, of the year. / Plants and animals can't live in a polar desert, but they can live in a tundra biome.

2. E.g. Its white fur helps it blend into the snow. / Its thick fur keeps it warm in the cold weather. / It has short ears and a short nose to reduce heat loss.

3. Any appropriate answer. Pupils' drawings should show an understanding of how polar bears stay warm, for example, by having a thick, oily coat and a layer of guard hairs over their soft undercoat fur.

4. Pupils may answer either way as long as they give sensible reasons for their choice. E.g. *Good:* The pipelines get oil to the oil companies. / We use oil to make plastics and fuel which we need. *Bad:* The pipelines affect wildlife (such as reindeer) and stop them from migrating. Pupils' explanations should expand one or more of the reasons they listed in the table e.g. *I think crude oil pipelines shouldn't be banned from the tundra, because we use oil for fuel and to make plastics which are very important in our daily lives. Unless the oil companies can access the oil, it can't be turned into things we need.*

Extra Activities

- Challenge pupils to build their own igloo using ice cubes or sugar cubes. If using ice cubes, suggest to pupils that they sprinkle a little salt on the ice cubes to help them stick together. If using sugar cubes, suggest PVA glue. Make sure they leave room for a door. Encourage pupils to think about how they will construct the igloo before they start. If their first attempt is not successful, encourage them to reflect briefly on what went wrong and how they can improve their next attempt.

- Provide pupils with one or more images of the Northern Lights. Ask them to create their own Northern Lights image using pastel chalks. Then ask pupils to draw trees or buildings onto black paper and cut them out. They can place these on top of their Northern Lights picture to create a silhouette effect.

- Pupils could test the thermal efficiency of their own coats. Ask pupils to fill two bottles with warm water and use a thermometer to record the temperature of the water in each. Place both bottles outside, one wrapped in a coat, and the other uncovered. Leave outside for 15-20 minutes, then record the temperature again. Which bottle of water remained the warmest? Pupils could repeat the experiment using different coats to see whose coat retains the most heat.

Discover & Learn Human and Physical Geography — Living Planet

Water Worlds

Study Book (pages 26-27)

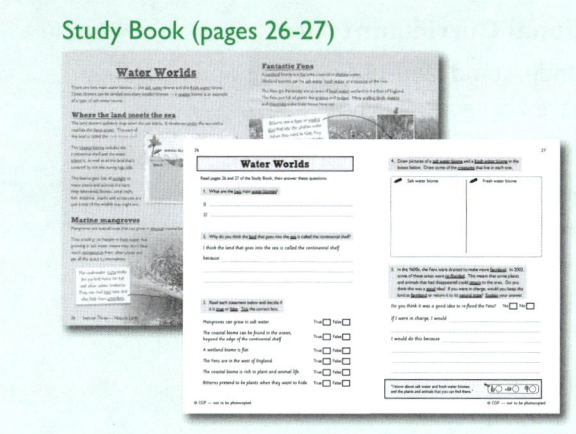

Activity Book (pages 26-27)

National Curriculum Aims
- Locate the world's countries, concentrating on their environmental regions and key human and physical characteristics.
- Describe and understand key aspects of physical geography, including climate zones and biomes.

Introduction

This topic covers some examples of fresh water and salt water biomes. Both types of water biome can be further divided into smaller biomes, for example estuaries, salt water wetlands and coral reefs.

Once pupils have read pages 26 and 27 of the Study Book, recap with the class what they have learned already about the oceans, including the three layers of the ocean and the creatures that live there, and how the oceans affect the climate.

Answers to Activity Book Questions

1. Fresh water biome and salt water biome.

2. E.g. *I think the land that goes into the sea is called the continental shelf because* it is the land at the edge of the continent. It looks like a shelf underwater, separating the coastal biome from the deep ocean.

3. True — False — True — False — True — True

4. Any appropriate answer. Pupils should draw on information from the Study Book. E.g. The salt water biome could include: seaweed, octopuses, sharks, sea horses, jellyfish, whales, dolphins. The fresh water biome could include: reeds and sedges, water lilies, fish, wading birds, otters.

5. Pupils may answer either way, as long as they give reasonable justification for their answer. E.g. Yes. *If I were in charge I would* return the land to its natural state. *I would do this because* I think it's important to protect the natural habitats of plants and animals so they don't become endangered or go extinct.
No. *If I were in charge I would* keep the land as farmland. *I would do this because* people need food and farmers need farmland to earn money.

Extra Activities

- Find a video about coral reefs aimed at KS2 pupils online and show it to the class. Encourage pupils to think about the diversity of life that exists in the coastal biome. How many different creatures in the video could they identify? Can they identify any predators? You could also ask them to discuss any environmental issues raised by the video, as a class or in small groups.

- Using the information from page 27 of the Study Book, ask pupils to create a poster to encourage people to consider their point of view regarding the re-flooding of the Fens. Encourage pupils to work on their persuasive skills by making their points as clearly and concisely as possible.

- Ask pupils to make their own coral reef — this can be done either individually or as a class project, using different coloured pipe cleaners, assorted colours of tissue paper, various colours of acetate sheet, sweet wrappers etc. Pupils could spread sand over a layer of PVA glue to represent the sea bed.

Making Changes

Study Book (pages 28-29)

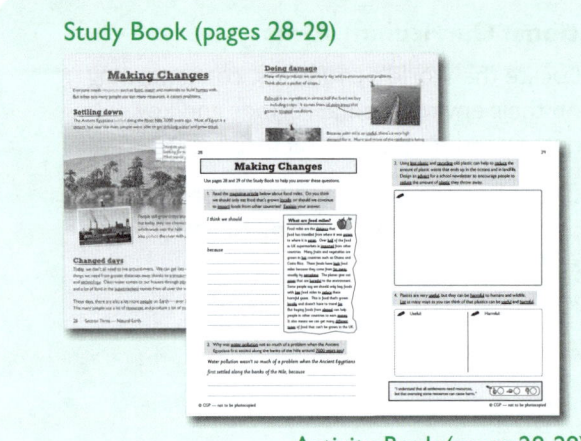

Activity Book (pages 28-29)

National Curriculum Aims
- Understand land-use patterns and how they have changed over time.
- Describe and understand key aspects of human geography, including types of settlement and land use, economic activity and the distribution of natural resources including food and water.

Introduction

This topic introduces pupils to some basic facts about the resources people need in a settlement. It also discusses the consequences of overuse or careless usage of certain resources, like plastic and palm oil.

Palm oil is the most widely used vegetable oil in the world. Recently, it has been under particular scrutiny for the impact its production has on the rainforest. However, campaigns against palm oil production are complicated by the fact that palm oil has a very high yield per square kilometre — higher than any other vegetable oil — which means that if we were to replace all palm oil with other vegetable oils, more land would have to be cleared to grow it. One solution could be closer regulation of palm oil plantations to reduce rainforest loss.

Answers to Activity Book Questions

1. E.g. *I think we should* keep importing food, *because* it gives us lots of choice in the supermarkets and helps farmers in other countries make money. / *I think we should* eat food grown locally, *because* it has low food miles, which would reduce the amount of harmful gases in the atmosphere.

2. E.g. *Water pollution wasn't so much of a problem when the Ancient Egyptians first settled along the banks of the Nile, because* there weren't any factories along the river to pollute the water / people weren't using harmful chemicals like pesticides and fertilisers that end up in the river.

3. Any appropriate answer. Pupils' posters should encourage people to reduce, reuse and recycle plastics.

4. E.g. *Useful:* making water bottles / making parts for computers, televisions and phones / sports equipment / pens and other school supplies / furniture / car parts / keeping food fresh.
   Harmful: plastics can hurt ocean animals that try to eat it or get stuck in it / plastic pollutes the water / plastic doesn't degrade so it fills up landfill sites.

Extra Activities

- As a class, using the information in the introduction above as a starting point, discuss the pros and cons of palm oil. Encourage pupils to think about sustainability when shopping with parents and consider more environmentally friendly products. Pupils could also create a display of items that use palm oil.

- Pupils could create a recycled display, for example, an image of a river flowing though a landscape, made entirely out of old straws, plastic lids, rubbish from break-time to convey they message that we need to cut down on the amount of waste we create.

- To expand on question 1 in the Activity Book, search online for a video aimed at KS2 pupils about food miles. Ask pupils to bring in empty food packets from home. Pupils can locate the countries of origin of the foods and mark them on a large wall map to show how far some of their own food has travelled.

Discover & Learn Human and Physical Geography — Living Planet

Biome Summary

Study Book (pages 14-29)

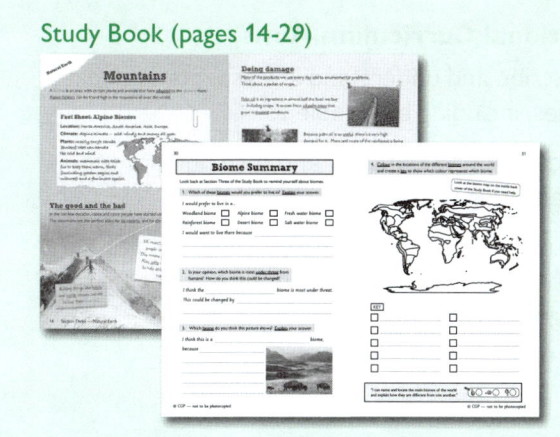

Activity Book (pages 30-31)

National Curriculum Aims

- Describe and understand key aspects of physical geography, including climate zones, biomes and vegetation belts.
- Locate the world's countries, concentrating on their environmental regions.
- Understand geographical similarities and differences through the study of physical geography.

Introduction

This topic allows pupils to bring together the knowledge they have gathered about the world's biomes by comparing and contrasting the various features of each biome.

Answers to Activity Book Questions

1. Pupils may choose any option as long as they give sensible reasons for their answer.

2. Any appropriate answer. E.g. *I think the* rainforest *biome is most under threat. This could be changed by* using less paper and recycling paper / reducing the amount of palm oil we use, so less rainforest is destroyed. *I think the* salt water *biome is most under threat. This could be changed by* recycling plastic waste and using less plastic, so that it doesn't end up the oceans where it can harm the wildlife.

3. E.g. *I think this is a* temperate grassland *biome, because* there are grasses and bushes, but no trees. There are bison, which live in temperate grasslands.

4. The map and the key should be coloured in so that each area of the map matches the correct biome in the key below.

Extra Activities

- Divide pupils into groups and give each group a biome to research. Each group could create a poster displaying the following information about their biome: location, temperature, precipitation, typical flora, typical fauna and any unique/interesting features. These posters could form a class display about biomes.

- Ask pupils to make a 'shoebox biome'. Taking a shoebox and placing it on its side, pupils can create a scene from a biome of their choosing inside. First, ask pupils to paint the background on what was the bottom of the shoebox, e.g. for an alpine biome, pupils could paint a blue sky and some mountains in the distance; for a tropical rainforest they could paint a cloudy sky with tall trees. Pupils can then add vegetation and ground cover, e.g. grass and short bushes for a grassland biome, or trees and fallen leaves for a deciduous forest. Pupils could either draw and cut out their ground cover on pieces of paper or card, or bring in pieces of plants or leaves. Finally, using pieces of card, pupils can draw and cut out a few examples of the animals that can be found in their chosen biome and place them in the display. Pupils should use a shoebox without a lid attached or any shiny coating that may make it difficult to colour or paint.

- Ask pupils to write an email or letter to organisations linked with environmental protection, e.g. WWF®, Greenpeace™ or the RSPB®, to gain more information about their work.

Discover & Learn Human and Physical Geography — Living Planet

Climate Change

Study Book (pages 30-31)

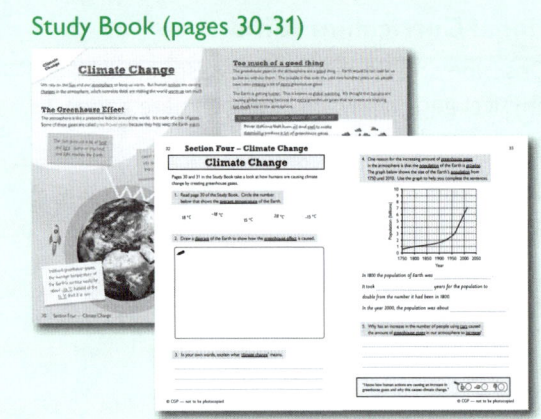

Activity Book (pages 32-33)

National Curriculum Aims

Describe and understand key aspects of climate zones, including being able to:

- describe and understand how humans are affecting the climate,
- understand how human activity, such as burning fossil fuels, is causing the release of greenhouse gases.

Introduction

This topic introduces pupils to greenhouse gases, the greenhouse effect and climate change. The amount of greenhouse gases, like carbon dioxide, in the Earth's atmosphere have varied greatly throughout history, with correlated changes in global temperatures. However, carbon dioxide levels are now increasing beyond the limits of natural changes in concentration that have been seen in the past.

The phrases "global warming" and "climate change" are often used interchangeably, but refer to different phenomena. Global warming refers to the long-term rise in the average global temperatures, driven by the increase in greenhouse gases in the atmosphere. Climate change is a broader term, referring to many changes in the global climate, such as an increased frequency of extreme weather events.

Answers to Activity Book Questions

1. 15 °C
2. An appropriate drawing that shows heat from the Sun reaching the Earth, radiating back out from the Earth towards space and then being reflected back to the Earth by greenhouse gases in the atmosphere.
3. E.g. climate change means a change in the world's temperature and weather that happens over a long time.
4. 1 billion — 125 (or any number between 120 and 130) — 6 billion
5. E.g. (most) cars burn fossil fuels. If there are more cars, more fuel gets burnt, so more greenhouse gas/carbon dioxide is produced.

Extra Activities

- Ask pupils to construct a diagram to show the relationships between humans, plants and animals. Ask them to explain why they are interdependent and need the others to survive. Pupils could be asked to predict what might happen to plants and animals as the Earth gets hotter, and what would happen to humans if certain plants or animals went extinct. They could present their predictions as flow diagrams.

- Ask pupils to list as many ways as they can think of that their lives contribute to carbon dioxide production. This could include transport, electricity, heating, food production and clothing production. For each thing they name, they could then try to come up with one way to reduce its contribution to their carbon footprint. Pupils could list their answers in the form of a table, with ways that they contribute to carbon dioxide production in one column and ways that they could reduce their contribution in a second column.

- As a class, have a debate or discussion about how humans are causing climate change, and what could be done about it. Before starting, pupils could be given some key words that they could use in their arguments (e.g. fossil fuels, greenhouse gases, greenhouse effect).

Effects of Climate Change

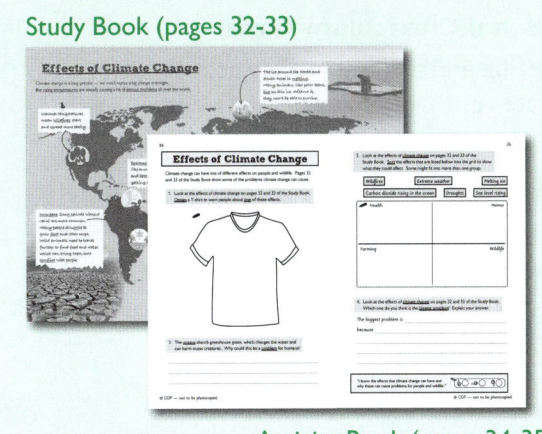

Study Book (pages 32-33)

Activity Book (pages 34-35)

National Curriculum Aims

Describe and understand key aspects of climate zones, including being able to:

- describe and understand how humans are affecting the climate,
- describe and understand the effects of climate change on people and wildlife around the world.

Introduction

This topic lets pupils explore the effects of climate change on both humans and wildlife. Some of the specific effects can be quite difficult to predict as it's hard to know exactly what is going to happen in the future. For example, predictions of the rise in sea level vary between 20 cm and 250 cm by 2100, depending on the different models and scenarios that are used. This uncertainty means that it's hard to know where and when climate change will have the biggest impacts on humans and natural habitats, but we'll all be affected.

Answers to Activity Book Questions

1. Any suitable design based on one of the seven effects shown on pages 32 and 33 of the Study Book.
2. E.g. there will be fewer fish in the sea, so people might have less to eat.
3. Health: Wildfires, Extreme weather, Droughts.
   Homes: Wildfires, Extreme weather, Sea level rising.
   Farming: Droughts, Extreme weather, Wildfires. (Sea level rising could also be included).
   Wildlife: Wildfires, Extreme weather, Melting ice, Carbon dioxide rising in the ocean, Droughts, Sea level rising.
4. Any answer with a suitable explanation. E.g. melting ice, because it is causing the sea levels to rise. This means that coastal or low-lying areas may be at risk of flooding. Some people could lose their homes. There may be less land to grow food, so some people might starve.

Extra Activities

- As a class, listen to the song "Mercy Mercy Me" by Marvin Gaye. Pupils can then discuss the lyrics and what they think the song is about. The class could also listen to "Love Song to the Earth" — a song released ahead of the UN climate conference in 2015. Pupils could then write their own song about climate change — they could use the tune of an existing song, such as a nursery rhyme, but write their own lyrics.

- In small groups, pupils can run an experiment to show the effects of melting sea ice and melting land ice. Each group has two waterproof boxes, and an island is placed in each tray (e.g. made from a smaller box or a stone). In the first box (land ice), water is poured to beneath the level of the island, and ice cubes are placed on the island. In the other box (sea ice), ice cubes are placed around the island, and water is added to beneath the level of the island. The depth of the water in each tray should be measured before and after the ice melts. Pupils can then discuss why there is a difference in sea level rise between the two trays.

- As a class or in pairs, get pupils to discuss which effect of climate change they think has the greatest impact and which has the least impact, and why.

Discover & Learn Human and Physical Geography — Living Planet

The Future

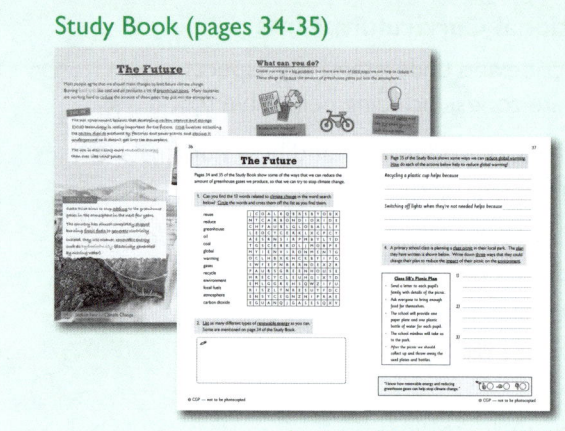

Study Book (pages 34-35)

Activity Book (pages 36-37)

National Curriculum Aims

Describe and understand key aspects of climate zones, including being able to:

- describe and understand how humans are affecting the climate,
- understand how changes in economic activities and the use of natural resources could reduce the release of greenhouse gases and slow climate change.

Introduction

This topic introduces pupils to the things that can be done to reduce greenhouse gas emissions and climate change. They may have awareness of some actions already, such as recycling waste at home.

Recycling reduces both the amount of waste that ends up in landfill and the energy needed to make products. For example, making a can from recycled aluminium requires around 95% less energy than it does to make it from new aluminium extracted from its ore.

Answers to Activity Book Questions

1. On the right are the locations of the words in the grid:

2. E.g. wind power, solar power, hydroelectricity, tidal power, biofuels/bio energy, geothermal energy.

3. *Recycling a plastic cup helps because* it means that less new plastic will need to be made, so less greenhouse gases will be produced by factories. *Switching off lights when they're not needed helps because* it means less electricity is used, so less oil and coal are burned in power plants to generate energy.

4. Any three from: e.g. send an e-mail rather than a letter / use reusable plates / ask pupils to bring a refillable water bottle each / walk to the park / recycle the bottles and plates instead of throwing them away.

Extra Activities

- Get pupils to design something that can be made with an item that would normally be thrown away (e.g. making a pen pot from a glass jar, or a musical instrument with empty yoghurt pots). Ask pupils to make their designs without using any new materials.

- Ask pupils to work as a class (or in smaller groups) to plan a presentation or assembly on how other pupils can cut down on their energy use. This could include talking about making sure things are switched off when you're not using them, recycling and any other ways they can think of that energy use could be reduced at school.

- Discuss with pupils the things that some countries, such as Costa Rica, are doing to tackle climate change. Ask them to suggest what other countries could do to cut their greenhouse gases (e.g. using renewable energy sources, recycling more). Pupils could then write letters to their MP or the government asking them to do more to tackle greenhouse gas emissions and climate change.

Discover & Learn Human and Physical Geography — Living Planet